Shaking the Heavens and Splitting the Earth

Chinese Air Force Employment Concepts in the 21st Century

Roger Cliff, John Fei, Jeff Hagen, Elizabeth Hague, Eric Heginbotham, John Stillion

Prepared for the United States Air Force

PROJECT AIR FORCE

The research described in this report was sponsored by the United States Air Force under Contract FA7014-06-C-0001. Further information may be obtained from the Strategic Planning Division, Directorate of Plans, Hq USAF.

Library of Congress Cataloging-in-Publication Data

Shaking the heavens and splitting the earth : Chinese air force employment concepts in the 21st century / Roger Cliff ... [et al].
    p. cm.
  Includes bibliographical references.
  ISBN 978-0-8330-4932-2 (pbk. : alk. paper)
    1. China. Zhongguo ren min jie fang jun. Kong jun. 2. Air forces—China. 3. Air power—China. I. Cliff, Roger. II. Title: Chinese air force employment concepts in the 21st century.

  UG635.C6S53 2011
  358.40951—dc22

                                                          2010037764

The RAND Corporation is a nonprofit institution that helps improve policy and decisionmaking through research and analysis. RAND's publications do not necessarily reflect the opinions of its research clients and sponsors.

RAND® is a registered trademark.

Cover image courtesy of AP Photos.

© Copyright 2011 RAND Corporation

Published 2011 by the RAND Corporation
1776 Main Street, P.O. Box 2138, Santa Monica, CA 90407-2138
1200 South Hayes Street, Arlington, VA 22202-5050
4570 Fifth Avenue, Suite 600, Pittsburgh, PA 15213-2665
RAND URL: http://www.rand.org/
To order RAND documents or to obtain additional information, contact
Distribution Services: Telephone: (310) 451-7002;
Fax: (310) 451-6915; Email: order@rand.org

Preface

China's air force is in the midst of a transformation. A decade ago, it was an antiquated service equipped almost exclusively with weapons based on 1950s-era Soviet designs and operated by personnel with questionable training according to outdated employment concepts. Today, the People's Liberation Army Air Force (PLAAF) appears to be on its way to becoming a modern, highly capable air force for the 21st century.

This monograph analyzes publications of the Chinese military, previously published Western studies of China's air force, and information available in published sources about current and future capabilities of the PLAAF. It describes the concepts for employing forces that the PLAAF is likely to implement in the future, analyzes how those concepts might be realized in a conflict over Taiwan, assesses the implications of China implementing these concepts, and develops recommendations about actions that should be taken in response. The book should be of interest to defense planners, analysts of China's military forces, policymakers, and anyone else interested in China's military modernization and its implications for the United States and Taiwan.

The book is the result of a project called "Chinese Air and Space Power," the purpose of which was to help the U.S. Air Force (USAF) understand how the Chinese military thinks about air and space power, how China might employ air and space power in a confrontation with the United States, and how the USAF can better counter Chinese doctrinal and operational concepts. The research reported here was sponsored by the Director of Air, Space and Information Operations, Headquarters Pacific Air Forces (PACAF A3/A5) and conducted within the Strategy and Doctrine Program of RAND Project AIR FORCE. It will be followed by a companion piece on Chinese space power and is part of an ongoing effort by Project AIR FORCE to assess the nature and implications of the growth in Chinese military power. The information in this monograph was current as of July 2009. Previous publications from this effort include the following:

- Evan S. Medeiros, *China's International Behavior: Activism, Opportunism, and Diversification*, Santa Monica, Calif.: RAND Corporation, MG-850-AF, 2009
- Evan S. Medeiros, Keith Crane, Eric Heginbotham, Norman D. Levin, Julia F. Lowell, Angel Rabasa, and Somi Seong, *Pacific Currents: The Responses of U.S. Allies and Security Partners in East Asia to China's Rise*, MG-736-AF, 2008
- Roger Cliff and David A. Shlapak, *U.S.-China Relations After Resolution of Taiwan's Status*, MG-567-AF, 2007
- Roger Cliff, Mark Burles, Michael S. Chase, Derek Eaton, and Kevin L. Pollpeter, *Entering the Dragon's Lair: Chinese Antiaccess Strategies and Their Implications for the United States*, MG-524-AF, 2007
- Evan S. Medeiros, Roger Cliff, Keith Crane, and James C. Mulvenon, *A New Direction for China's Defense Industry*, MG-334-AF, 2005
- Keith Crane, Roger Cliff, Evan Medeiros, James C. Mulvenon, and William H. Overholt, *Modernizing China's Military: Opportunities and Constraints*, MG-260-1-AF, 2005
- Kevin Pollpeter, *U.S.-China Security Management: Assessing the Military-to-Military Relationship*, MG-143-AF, 2004
- Zalmay Khalilzad, David T. Orletsky, Jonathan D. Pollack, Kevin L. Pollpeter, Angel Rabasa, David A. Shlapak, Abram N. Shulsky, and Ashley J. Tellis, *The United States and Asia: Toward a New U.S. Strategy and Force Posture*, MR-1315-AF, 2001
- Roger Cliff, *The Military Potential of China's Commercial Technology*, MR-1292-AF, 2001

- Erica Strecker Downs, *China's Quest for Energy Security*, MR-1244-AF, 2000
- Richard Sokolsky, Angel Rabasa, and C. Richard Neu, *The Role of Southeast Asia in U.S. Strategy Toward China*, MR-1170-AF, 2000
- Abram N. Shulsky, *Deterrence Theory and Chinese Behavior*, MR-1161-AF, 2000
- Mark Burles and Abram N. Shulsky, *Patterns in China's Use of Force: Evidence from History and Doctrinal Writings*, MR-1160-AF, 2000
- Michael D. Swaine and Ashley J. Tellis, *Interpreting China's Grand Strategy: Past, Present, and Future*, MR-1121-AF, 2000
- Daniel Byman and Roger Cliff, *China's Arms Sales: Motivations and Implications*, MR-1119-AF, 1999
- Zalmay Khalilzad, Abram N. Shulsky, Daniel Byman, Roger Cliff, David T. Orletsky, David A. Shlapak, and Ashley J. Tellis, *The United States and a Rising China: Strategic and Military Implications*, MR-1082-AF, 1999
- Mark Burles, *Chinese Policy Toward Russia and the Central Asian Republics*, MR-1045-AF, 1999.

RAND Project AIR FORCE

RAND Project AIR FORCE (PAF), a division of the RAND Corporation, is the U.S. Air Force's federally funded research and development center for studies and analyses. PAF provides the Air Force with independent analyses of policy alternatives affecting the development, employment, combat readiness, and support of current and future aerospace forces. Research is conducted in four programs: Force Modernization and Employment; Manpower, Personnel, and Training; Resource Management; and Strategy and Doctrine.

Additional information about PAF is available on our website: http://www.rand.org/paf/

Contents

Figures

Summary

China's air force is in the midst of a transformation. A decade ago, it was an antiquated service equipped almost exclusively with weapons based on 1950s-era Soviet designs and operated by personnel with questionable training according to outdated employment concepts. Today, the PLAAF appears to be on its way to becoming a modern, highly capable air force for the 21st century. This monograph describes the concepts that the PLAAF is likely to implement in the future for employing its aviation, surface-to-air missile (SAM), antiaircraft artillery (AAA), and airborne forces; analyzes how those concepts might be implemented in specific operational situations; assesses the implications for the USAF of the PLAAF implementing these concepts, given the capabilities it currently possesses or may acquire in the future; and develops recommendations for the USAF about actions it should take in response.

Research Approach

The overall approach of the study from which this monograph results was to analyze publications of the Chinese military, as well as previously published Western analyses of China's air force, for information on how the PLAAF intends to employ its forces in the event of a future conflict; combine these findings with information available in published sources about current and future capabilities of the PLAAF to assess how those general principles might be implemented in specific potential combat operations; and use a combination of expert judgment and quantitative analysis to identify implications and potential responses for the USAF. Most of the Chinese sources used in this study have not been translated into English, and all were read in the original Chinese to avoid the mistranslations, inconsistent use of terms, and other problems associated with most translations of Chinese military publications. The Chinese military publications used in this study were largely official reference books or textbooks used by China's military, collectively referred to as the People's Liberation Army (PLA). They do not necessarily reflect actual current practice, however, but rather appear to represent the views of the PLA, the PLAAF, and Chinese officers and analysts about how China's air forces *ought* to be employed, and thus can be regarded as a description of how the PLAAF aspires to operate in the future more than a documentation of how it is operated today (see pp. 4–10).

PLAAF Organization

Before discussing PLAAF employment concepts, it is useful to examine how Chinese airpower fits into the overall structure of the Chinese armed forces. The PLA consists of the PLA Army, the PLA Navy (PLAN), the PLAAF, and China's strategic rocket forces, known as the Second Artillery Force. For peacetime operations, China is divided up into seven Military Regions (MRs) (in protocol order): the Shenyang, Beijing, Lanzhou, Nanjing, Guangzhou, Jinan, and Chengdu MRs (see Figure 2.1 in Chapter Two). The commander of each MR—which, to date, has always been an army officer—has control over all PLA Army units, as well as all military operations, in his or her MR. During peacetime, however, the Chinese navy, air force, and Second Artillery Force are responsible for operational command, training, and other administrative and management issues of their respective forces in each MR. In the event of a war, a theater command would be established with operational command of all (conventional) military units within one or more MRs (see pp. 14–15).

China's aerospace power is contained in all four services of the PLA. In addition to the PLAAF, the PLA Army operates air defense

(SAM and AAA) and aviation (helicopter) forces; the PLAN has its own aviation forces, shore-based AAA, and shipboard SAM and AAA systems; and the Second Artillery Force operates conventional surface-to-surface missiles (SSMs). The PLAAF, moreover, comprises four combat branches: aviation, SAMs, AAA, and airborne (see pp. 15–27).

Key Employment Concepts and Principles

Official Chinese military publications define *airpower* as an over-all term for aviation forces belonging to air forces, navies, air defense forces (such as the Russian ProtivoVozdushnaya Oborona [Anti-Air Defense], or PVO), ground forces, and special operations forces. In joint operations, airpower is said to be used for high-speed, in-depth strikes against key targets and to be used first and throughout campaigns to seize control of the skies in support of broader campaign objectives. It also is used defensively to protect the ability of an air force to conduct air operations by defending air bases, air defense positions, and radar sites, as well as to protect ground and naval operations (see pp. 54–56).

PLA publications assert that the struggle for dominance of the battlefield will increasingly consist of an integrated struggle for air, space, information, and electromagnetic (and even computer network) superiority. Acquiring air superiority is considered a prerequisite in a variety of operations involving all services. By obtaining air superiority, one can restrict enemy air, air defense, and ground forces' operational movements while ensuring that one's own ground and navy forces have effective cover from the air to carry out their operations. Like the USAF, however, the PLA does not assert that achieving absolute air superiority in all stages of combat and across all battlefields or theaters is necessary. Instead, it aims to achieve enough air superiority to achieve its campaign or tactical objectives. Presumably because of reservations about its ability to defeat a qualitatively superior opponent, such as the United States, in the air, the PLA places primary emphasis on achieving air superiority by attacking the enemy on the ground and water: forces, equipment, bases, and launch pads used for air raids. Especially at the beginning of a war, the PLA will endeavor to attack enemy air bases, ballistic missile bases, aircraft carriers, and warships equipped with land-attack cruise missiles before enemy aircraft can take off or other forms of enemy air strike can be carried out. Another means of achieving air superiority will be to carry out attacks to destroy and suppress ground-based air defense systems and air defense command systems. In addition, defensive operations will be an important component of air superiority throughout a campaign (see pp. 56–60).

In future warfare, space superiority is expected to be crucial for controlling the ground, naval, and air battlefields. To gain space superiority, offensive and defensive weapon systems will be deployed on the ground, air, sea, and space. Space control operations are said to include space information warfare, "space blockade warfare," "space orbit attack warfare," space-defense warfare, and space-to-land attacks (see pp. 60–61).

In struggles for information superiority, the goal is to control information on the battlefield, allowing the battlefield to be transparent to one's own side but opaque to the enemy. Methods for achieving information superiority include achieving electromagnetic superiority through electronic interference; achieving network superiority through network attacks; using firepower to destroy the enemy's information systems; and achieving "psychological control" (see pp. 61–62).

While acquiring electromagnetic superiority is described as a subset of acquiring information superiority, it is treated as a distinct operation in PLA publications. Methods for obtaining electromagnetic superiority are said to include electronic attack and electronic defense. In electronic attack, soft kill measures include electronic interference and electronic deception. Hard kill measures are said to include antiradiation destruction, electromagnetic weapon attack, firepower destruction, and attacks against the enemy's electronic installations and systems. Electronic defense is simply defending against enemy electronic and firepower attacks. The primary targets of electronic warfare (EW) are said to include command, control, communications, and intelligence systems (see pp. 61–64).

PLAAF publications describe three major types of air combat operation: air-to-air combat, air-to-surface combat, and surface-to-air combat. Air-to-air operations are an area of traditional emphasis for the PLAAF, but the PLAAF seems to be moving away from emphasizing air-to-air operations and toward emphasizing operations to gain air superiority by attacking enemy airfields and controlling the enemy on the ground before resorting to fighting the enemy in the air. Air-to-surface operations are considered more effective, less costly, and less reactive than air-to-air operations (see pp. 65–78).

Campaign-Specific Employment Concepts

Chinese military publications identify four types of air force campaigns: air offensive campaigns, air defense campaigns, air blockade campaigns, and airborne campaigns. These can be either air force–only campaigns or, more frequently, air force–led joint campaigns that incorporate other services. These air force campaigns can also be part of broader joint campaigns, such as an island-landing campaign or joint blockade campaign. In most air operations, a great deal of emphasis is placed on surprise, camouflage, use of tactics, meticulous planning, and strikes against critical targets (see p. 85).

An air offensive campaign can include one or more of several objectives: obtaining air superiority; destroying key enemy political, military, and economic targets; destroying the enemy's transportation and logistic supply system; and destroying the enemy's massed forces to isolate the battlefield and facilitate PLA ground and maritime operations. Obtaining air superiority is needed in order to conduct air strikes against targets, but the ultimate objective of an air offensive campaign is to strike political, economic, and military targets. Several types of combat groups are involved in air offensive campaigns: a strike group, a suppression group, a cover group, a support group, an air defense group, and an operational reserve. An offensive air campaign is said to consist of four tasks: conducting information operations, penetrating enemy defenses, conducting air strikes, and resisting enemy counterattacks. The last of these is conducted throughout the campaign. The others are generally initiated sequentially, beginning with information operations (see pp. 89–113).

A textbook on military operations would list three primary missions for air defense campaigns: protecting the capital against air attack, protecting other important targets within the theater, and seizing and holding air superiority. Air defense campaigns, according to Chinese military writings, can be national in scope or can be confined to a particular theater. Depending on the circumstances, the entire air effort in a given war could be defensive; a single phase could be defensive; or, in the case of a geographically wide-ranging conflict, some theaters could be defensive, while some are offensive. In a war over Taiwan, for example, the PLA might conduct an offensive air campaign in the area opposite Taiwan while preparing for air defense campaigns to the north and south in anticipation of possible retaliation or counterattack by U.S. forces. Air defense campaigns are said to entail three types of operations: resistance, counterattack, and close protection. *Resistance operations* are actions to intercept, disrupt, and destroy attacking aircraft. *Counterattack operations* are attacks on enemy air bases (including aircraft carriers). *Close protection operations* are passive defense measures, such as fortification, concealment, camouflage, and mobility. China's overall approach to air defense is to combine the early interception of enemy attacks with full-depth, layered resistance to protect targets and forces while gradually increasing the tempo of counterattacks on enemy bases (see pp. 118–139).

Air blockade campaigns are operations to prevent an adversary from conducting air operations and to cut off its economic and military links with the outside world. Some Chinese sources describe them as simply a special variety of air offensive campaign, but most authoritative sources regard them as a distinct type of campaign. They will usually be conducted as part of a broader joint blockade campaign but can be implemented as an independent air force campaign. Air blockade campaigns are regarded as having a strong political nature, being long in duration, and requiring a high level of command and control. Typically, an air blockade campaign will entail the establishment of one or more no-fly zones surrounded by several aerial surveillance zones. Actions conducted as part of an air blockade campaign will include information operations, flight prohibition operations, interdiction of maritime and ground traffic, strikes against the enemy's counterblockade system, and air defense operations (see pp. 145–161).

Unlike the U.S. armed forces, the PLA's paratroops belong to its air force; therefore, an airborne campaign in the PLA is an air force campaign, not a joint campaign. Airborne campaigns are regarded as resource-intensive and difficult. For an airborne campaign to be carried out, information and air superiority must be seized (at least locally) and firepower preparation around the landing zone must be carried out. Then, air corridors to the landing zone must be opened up and kept clear, and enemy land-based air defenses near the landing zone must be suppressed while airborne forces are flown to the landing zone. Once they have landed, the airborne forces must clear and secure a base for receiving additional forces and supplies, including, if they landed on or near an airfield, seizing the airfield and bringing it to operational readiness. Meanwhile, friendly air and missile forces will suppress and interdict nearby enemy ground forces. Finally, the air-landed forces can initiate ground operations (see pp. 165–177).

Although any of these four types of air force campaigns can be conducted as an independent single-service campaign, they are more likely to be conducted as part of a broader joint campaign, such as an island-landing campaign or a joint blockade campaign. Even if an air force campaign is conducted as an independent, single-service campaign, moreover, other services, particularly the PLAN and the Second Artillery, are likely to be involved in supporting roles. For example, conventional missiles of the Second Artillery will play a key role in air offensive campaigns, counterattack operations of air defense campaigns, and providing firepower support for airborne campaigns. Similarly, the PLAN has responsibility for defending certain sectors of China's airspace and would be the service responsible for conducting counterattacks against air attacks launched from aircraft carriers and, thus, would likely play an important role in an air defense campaign. The PLAN is also responsible for providing air defense for surface naval forces, including, presumably, a Taiwan-bound invasion force. Little information appears to be available in published Chinese sources, however, on how PLAAF and PLAN aviation and SAM forces would interoperate when conducting air operations—a potentially significant challenge, particularly given the huge engagement envelopes (150 km or more) of the land-based and ship-based SAMs the PLAAF and PLAN have begun acquiring. Conversely, the PLAAF appears to have no naval strike mission or capability, meaning that naval strike operations are the sole responsibility of the relatively small and less-capable PLAN aviation forces (along with, possibly in the future, the Second Artillery, if it acquires an antiship ballistic missile capability) (see pp. 179–186).

Implications and Recommendations

By 2015 or so, the weapon systems and platforms China is acquiring will potentially enable it to effectively implement the four types of air force campaigns described in the previous section. The significant numbers of modern fighter aircraft and SAMs, as well as the long-range early warning radars and secure data and voice communication links China is likely to have by 2015, for example, coupled with the hardening and camouflage measures China has already taken, would make a Chinese air defense campaign, if conducted according to the principles described in Chinese military publications, highly challenging for U.S. air forces. Similarly, those same modern fighters, along with ground-launched conventional ballistic and cruise missiles, cruise missile–carrying medium bombers, and aerial refueling aircraft, will enable China to conduct offensive operations far into the western Pacific. Whether China will actually be able to fully exploit its air force doctrine and capabilities, however, is less clear. Much will depend on the quality of the training and leadership of China's air force, and it should be pointed out that the PLAAF last engaged in major combat operations in the Jinmen campaign of 1958, more than 50 years ago (see pp. 187–223).

The concepts and capabilities described in this monograph have a number of implications for the United States. First, if the United States intervenes in a conflict between the People's Republic of China (PRC) and Taiwan, it should expect attacks on its forces and facilities in the western Pacific, including those in Japan. Even in peacetime, therefore, the United States should take steps to prevent China from collecting information on military and sensitive civilian information systems or on U.S. early warning, command-and-control, SAM, and other sensors and communication systems. Similarly, U.S. forces should also ensure, to the maximum extent practical, that their information systems are protected from network intrusions or denial-of-service attacks while planning and training for the possibility that some of these systems will fail or be compromised in the event of an actual conflict. During such a conflict, the U.S. armed forces should prepare to deal with electronic jamming on a scale larger than it has seen in any conflict since the end of the Cold War. U.S. intelligence collectors should also expect extensive efforts to deceive them about the locations and posture of Chinese forces both prior to and during a conflict (see pp. 237–238).

Once the conflict begins, the United States should accept the likelihood that the runways of Okinawa's military airfields will be rendered at least temporarily unusable and that many or most unsheltered aircraft will be damaged or destroyed in the initial salvo of ballistic missiles, with sheltered aircraft, fuel storage and distribution facilities, and repair and maintenance facilities subject to follow-on attacks by cruise missiles and manned aircraft with precision-guided munitions (PGMs). One set of responses to this challenge would be to increase the number of missile defense systems on Okinawa; to build shelters capable of protecting all aircraft to be based on Okinawa; to harden runways, fuel, and repair facilities; to increase rapid runway repair capabilities; and to deploy mobile point-defense systems, such as the U.S. Army's Surface-Launched Advanced Medium-Range Air-to-Air Missile (SLAMRAAM), to defend Okinawa's air bases. If even vague indications are received that China might be planning to use force somewhere in East Asia, the United States should begin parking aircraft in shelters when not in use, begin keeping early warning and interceptor aircraft continuously airborne, and regularly relocate its SAM batteries to unpredictable sites (see pp. 238–239).

An alternative approach would be to keep relatively few combat aircraft on Okinawa in the event of a crisis over Taiwan and instead deploy the bulk of U.S. land-based air forces to several more-distant bases in Japan and elsewhere in the western Pacific. Even more-distant bases should not be regarded as sanctuaries, however, so the United States should nonetheless deploy active missile defenses, construct aircraft shelters, harden runways and facilities, and increase rapid runway repair capabilities at these bases. In either case, the USAF will need to continue to invest in fighter aircraft technology and pilot skill to ensure that it maintains its advantage in the face of rapid Chinese improvements in these areas (see pp. 239–240).

An alternative, or supplement, to fighter operations would be larger aircraft capable of carrying large numbers (e.g., 20 or more) of extremely long-range (e.g., 200 nm) air-to-air missiles. A supersonic bomber, such as the B-1, would be one possibility for providing this capability, as would be a stealthy aircraft like those that were considered for the USAF's now-canceled Next-Generation Bomber program. The missiles themselves could potentially be derivatives of existing airframes, such as those of the Patriot MIM-104 or SM-2ER RIM-67, perhaps coupled with a smaller second stage for the terminal engagement (see pp. 240–241).

In addition to improving its capabilities to defend Taiwan's airspace, the USAF should also examine ways to improve its capabilities to conduct offensive operations against China, as it may be that the most effective way to defeat China's air force in a conflict over Taiwan would be to attack China's aircraft while they were on the ground. The USAF's stealthy B-2 bomber can potentially be used to conduct such attacks, and, if a new-generation bomber becomes available, it will be able to provide this capability as well. An alternative to bombers penetrating into China's territory would be a long-range, stealthy cruise missile that could be launched at standoff ranges from bombers that the USAF possesses in larger numbers than the B-2. The stealthy Joint Air-to-Surface Standoff Missile–Extended Range (JASSM-ER) launched from B-1s or B-52s might be able to play this role for targets up to 300 nm inland. To reach targets further inland, a longer-range stealthy cruise missile would be needed, such as, if feasible, an Advanced Cruise Missile converted to carry a conventional warhead (see pp. 241–243).

In a conflict over Taiwan, the capabilities of Taiwan's armed forces would also be critical to the outcome, even if the United States intervened on a large scale. The longer Taiwan is able to deny the PRC air superiority over Taiwan, the more combat power the United States will be able to bring to the defense of Taiwan and the better the chances of a successful defense of the island. Defending Taiwan against air attack is feasible if Taiwan makes systematic, sustained, and carefully chosen investments.

Like the United States, therefore, Taiwan should take steps to prevent China from collecting information on military and sensitive civilian information systems or on Taiwan's early warning, command-and-control, SAM, and other sensors and communication systems. Moreover, in the event that a Chinese attack was planned, Taiwan's intelligence collectors should expect extensive efforts to deceive them about the locations and posture of Chinese forces. Taiwan's forces should also ensure, to the maximum extent practical, that their information systems are protected from network intrusions or denial-of-service attacks, and plan and train for the possibility that some of these systems would fail or be compromised in a conflict with the PRC. Once a Chinese offensive air campaign is under way, Taiwan should be prepared to deal with massive electronic jamming (see pp. 243–244).

It is not feasible for Taiwan to acquire enough missile defense systems to protect it against the simultaneous arrival of the number of ballistic missile warheads China is likely to fire at Taiwan in a conflict, though additional missile defenses, such as the six PAC-3 batteries Taiwan plans to acquire, will have some utility by increasing the number of ballistic missiles China would have to launch to be certain of putting out of action the runways at all of Taiwan's military airfields. For Taiwan's PAC-3 and PAC-2 systems to be effective, however, they must be relocated on a regular basis to unpredictable locations (see p. 244).

At least as important as to the defense of Taiwan, and possibly more cost-effective than active missile defenses, are passive defense measures, such as building shelters to protect Taiwan's combat aircraft from ballistic missile attack; hardening runways, fuel, and repair facilities; and increasing rapid runway repair capabilities at Taiwan's air bases. Ideally, the number of shelters would be several times the number of Taiwan's combat aircraft, with each aircraft randomly assigned to one of several different shelters every time it returned to base. Mobile point-defense systems, such as SLAMRAAM, could help defend Taiwan's air bases and other key targets. Finally, even if hostilities have not actually occurred, if there are indications that China might use force against Taiwan, as many aircraft as possible should be maintained aloft (see pp. 244–245).

Taiwan's defenders should expect the PRC's cruise missiles and aircraft to approach Taiwan not on a direct line from their launch points but from all directions, including the north, south, and east, and to make use of low altitude and terrain masking to disguise their approach. The attacking aircraft and missiles should be expected to focus their attacks first on Taiwan's own air and missile capabilities. An airborne landing, if attempted, would most likely occur in a lightly defended location in an area where the PRC could ensure continuous air superiority between the point of embarkation and the landing zone (see p. 245).

From what we find in Chinese military publications, Taiwan should also expect attacks on government, water, and electric installations and, if a prolonged campaign is expected, on key economic targets. Mitigating actions should be taken, such as ensuring that backup installations exist and evacuating government facilities, if there are indications that China might use force against Taiwan (see p. 245).

Acknowledgments

We would like to express our gratitude to a number of people who assisted in this study and helped with the preparation of the monograph. Kenneth W. Allen, then of CNA and now of Defense Group International, generously provided guidance and suggestions and the loan of several Chinese books used in this study. Jason Kelly and Max Woodworth, both then of RAND, checked the consistency of numerous passages with the original Chinese sources and provided revised text when needed. Kristin Smith, also then of RAND, formatted the draft and performed the time-consuming and tedious tasks of compiling the bibliography and glossary and ensuring that all citations are in the proper format. Michael S. Chase of the U.S. Naval War College and Forrest Morgan of RAND provided valuable and detailed reviews, while Kenneth Allen and David Yang (also of RAND) provided numerous valuable comments and corrections on various drafts. Meagan Smith of RAND reformatted the draft after the initial version was revised. Then–Strategy and Doctrine program director David Ochmanek supplied valuable advice, support, and great patience. Jane Siegel of PAF and Peter Hoffman, Lisa Bernard, Lauren Skrabala, Sandra Petitjean, Kimbria McCarty, Carol Earnest, Jocelyn Lofstrom, and Stephan Kistler of RAND Publications and Creative Services oversaw the editing, formatting, and graphic design of the book.

Abbreviations

A3/A5	Director of Air, Space and Information Operations
AAA	antiaircraft artillery
ACM	Advanced Cruise Missile
AFDD	Air Force Doctrine Document
ALCM	air-launched cruise missile
AMS	Academy of Military Science
AWACS	airborne warning and control system
BDA	battle damage assessment
C4I	command, control, communications, computers, and intelligence
C4ISR	command, control, communications, computers, intelligence, surveillance, and reconnaissance
CAP	combat air patrol
CAS	close air support
CMC	Central Military Commission
DCA	defensive counterair
ELINT	electronic intelligence

EW	electronic warfare
FAE	fuel-air explosive
GCI	ground control intercept
GPS	Global Positioning System
ICBM	intercontinental ballistic missile
IFF	identification, friend or foe
IRSTS	infrared search and track system
ISR	intelligence, surveillance, and reconnaissance
JASSM-ER	Joint Air-to-Surface Standoff Missile–Extended Range
JDAM	Joint Direct Attack Munition
JP	Joint Publication
JSTARS	Joint Surveillance Target Attack Radar System
LDHD	low density, high demand
MR	Military Region
MRAF	Military Region Air Force
NATO	North Atlantic Treaty Organization
NGB	next-generation bomber
OCA	offensive counterair
PACAF	Pacific Air Forces
PGM	precision-guided munition
PLA	People's Liberation Army
PLAAF	People's Liberation Army Air Force
PLAN	People's Liberation Army Navy

PRC	People's Republic of China
PVO	ProtivoVozdushnaya Oborona
SAM	surface-to-air missile
SEAD	suppression of enemy air defense
SLAMRAAM	Surface-Launched Advanced Medium-Range Air-to-Air Missile
SOF	special operations force
SOJ	standoff jammer
SSM	surface-to-surface missile
TBM	theater ballistic missile
UAV	unmanned aerial vehicle
USAF	U.S. Air Force
USN	U.S. Navy

CHAPTER ONE

Introduction

China's air force is in the midst of a transformation. A decade ago, it was an antiquated service equipped almost exclusively with weapons based on 1950s-era Soviet designs and operated by personnel with questionable training according to outdated employment concepts. Today, the People's Liberation Army Air Force (PLAAF) appears to be on its way to becoming a modern, highly capable air force for the 21st century. In 1999, for example, the PLAAF operated about 3,500 combat aircraft. Of these, the vast majority were based on the Soviet MiG-19 and MiG-21 fighter aircraft, both of which first flew in the 1950s. In 1999, China possessed fewer than 100 modern fighter aircraft—all Su-27s purchased from Russia—and only 100 or so H-6 medium bombers (also based on a 1950s Soviet design). Only China's Su-27 aircraft carried beyond–visual range air-to-air missiles, and China's air force possessed no precision-guided munitions (PGMs). The pilots of all aircraft types probably received fewer than 100 hours of flight time per year and rarely flew at night, at low level, in bad weather, or over water. The situation for the PLAAF's surface-to-air missile (SAM) and antiaircraft artillery (AAA) forces was similar to that for its air forces.[1]

---

[1] International Institute for Strategic Studies, *The Military Balance 1999/2000*, London: Oxford University Press, 1999, p. 188; Evan S. Medeiros, Roger Cliff, Keith Crane, and James C. Mulvenon, *A New Direction for China's Defense Industry*, Santa Monica, Calif.: RAND Corporation, MG-334-AF, 2005, pp. 160, 163–164; Kenneth W. Allen, "PLAAF Modernization: An Assessment," in James R. Lilley and Chuck Downs, eds., *Crisis in the Taiwan Strait*, Washington, D.C.: National Defense University Press, September 1997, pp. 217–248.

The picture today is quite different. As of 2010, the PLAAF has retired many of its older aircraft and is operating more than 300 modern fighter aircraft, with more in production. These include Russian-designed Su-27s and Su-30s but also China's own domestically developed J-10, which is assessed to be comparable in capability to the U.S. F-16. Many PLAAF fighters now carry beyond–visual range air-to-air missiles and PGMs, and the PLAAF possesses a first-generation air-launched cruise missile (ALCM), carried on the H-6 medium bomber. Chinese pilots now average well over 100 hours of flight time per year, and the pilots of the most-advanced fighters are believed to receive close to 200 hours per year. China is experimenting with domestically produced airborne warning and control system (AWACS) aircraft, and PLAAF aircraft now routinely operate at low level, over water, in bad weather, and at night (sometimes all at once). Meanwhile, the PLAAF's SAM forces have purchased the modern S-300 series of SAMs (North Atlantic Treaty Organization [NATO] designators SA-10 and SA-20) from Russia and have fielded a domestic system (the HQ-9) of comparable capability.[2]

Based on recent trends, these changes are likely to accelerate in the future, so that, within another decade, the capabilities of China's air force could begin to approach those of the U.S. Air Force (USAF) today. USAF capabilities will continue to improve as well, of course, so that it will still enjoy a significant qualitative advantage, but a conflict with China might not be the lopsided contest it likely would have

---

[2] International Institute for Strategic Studies, *The Military Balance 2010*, London: Routledge, 2010, p. 404; "Fei Teng Guided Bombs (FT-1, FT-2, FT-3, FT-5)," *Jane's Air-Launched Weapons*, January 13, 2010; "LT-2 Laser Guided Bomb," *Jane's Air-Launched Weapons*, July 27, 2007; "YJ-91, KR-1 (Kh-31P)," *Jane's Air-Launched Weapons*, October 15, 2007; "Chinese Signals Intelligence (SIGINT) Air Vehicles," *Jane's Electronic Mission Aircraft*, January 8, 2007; "CAC J-10," *Jane's All the World's Aircraft*, April 14, 2010; "Chinese Laser-Guided Bombs (LGBs)," *Jane's Air-Launched Weapons*, January 17, 2008; "HQ-9/-15, and RF-9 (HHQ-9 and S-300)," *Jane's Strategic Weapon Systems*, January 28, 2008; "KD-63 (Kong Di-63)," *Jane's Air-Launched Weapons*, January 25, 2008; "LS-6 Glide Bomb," *Jane's Air-Launched Weapons*, January 17, 2008; "SAC Y-8/Y-9 (Special Mission Versions)," *Jane's All the World's Aircraft*, November 12, 2009; "SD-10 (PL-12)," *Jane's Air-Launched Weapons*, January 22, 2008; Office of the Secretary of Defense, *Annual Report to Congress: Military Power of the People's Republic of China 2008*, Washington, D.C.: U.S. Department of Defense, 2008.

been in the late 1990s. And, even today, the emerging capabilities of the PLAAF are such that, combined with the geographic and other advantages China would enjoy in the most likely conflict scenario—a war over Taiwan—the USAF could find itself challenged in its ability to achieve air dominance over its adversary, a prospect that the USAF has not had to seriously consider for nearly two decades.

This is a key time, therefore, to develop a better understanding of how China's air force might actually employ the capabilities it is acquiring. Much reporting and analysis on China's military focuses on the weapons and other systems that it is acquiring.[3] Less often, it examines such dimensions as organization, training, leadership, and personnel.[4] Such reports help illuminate the current and future capabilities of China's military, but they do not describe how those capabilities might be employed and therefore what specific challenges the USAF is likely to face in the event of a conflict with China. Although some valuable work has been done in this area,[5] the study from which the present monograph results was initiated based on the observation that

---

[3] For example, see Office of the Secretary of Defense, 2008; Richard D. Fisher Jr., "PLA Air Force Equipment Trends," in Stephen J. Flanagan and Michael E. Marti, eds., *The People's Liberation Army and China in Transition*, Washington, D.C.: National Defense University Press, August 2003, pp. 139–176.

[4] For example, see Kenneth W. Allen, "PLA Air Force Organization," in James C. Mulvenon and Andrew N. D. Yang, eds., *The People's Liberation Army as Organization: Reference Volume v1.0*, Santa Monica, Calif.: RAND Corporation, CF-182-NSRD, 2002, pp. 346–457; Kenneth W. Allen, "PLA Air Force, 1949–2002: Overview and Lessons Learned," in Laurie Burkitt, Andrew Scobell, and Larry M. Wortzel, eds., *The Lessons of History: The Chinese People's Liberation Army at 75*, Carlisle, Pa.: Strategic Studies Institute, July 2003, pp. 89–156; Dennis J. Blasko, *The Chinese Army Today: Tradition and Transformation for the 21st Century*, London: Routledge, 2006; Bernard D. Cole, *The Great Wall at Sea: China's Navy Enters the Twenty-First Century*, Annapolis, Md.: Naval Institute Press, 2001; Kevin Lanzit and Kenneth W. Allen, "Right-Sizing the PLA Air Force: New Operational Concepts Define a Smaller, More Capable Force," in Roy Kamphausen and Andrew Scobell, eds., *Right-Sizing the People's Liberation Army: Exploring the Contours of China's Military*, Carlisle, Pa.: Strategic Studies Institute, U.S. Army War College, September 2007, pp. 437–478; Office of Naval Intelligence, *China's Navy 2007*, Washington, D.C.: Department of the Navy, 2007.

[5] For example, see James C. Mulvenon and David Michael Finkelstein, eds., *China's Revolution in Doctrinal Affairs: Emerging Trends in the Operational Art of the Chinese People's Liberation Army*, Alexandria, Va.: CNA Corporation, December 2005.

the amount that has been published on this topic is small compared to the amount of information that is available in the open publications of China's military. The goals of this study were to identify the concepts that the PLAAF is likely to implement in the future for employing its aviation, SAM, AAA, and airborne forces; to analyze how those concepts might be implemented in specific operational situations; to assess the implications for the USAF of the PLAAF implementing these concepts, given the capabilities it currently possesses or may acquire in the future; and to develop recommendations for the USAF about actions it should take in response to those implications.

Methodology

The overall approach of the study from which this monograph results was to analyze publications of the Chinese military and previous Western analyses for information on how the PLAAF intends to employ its forces in the event of a future conflict; combine these findings with information available in published sources about current and future capabilities of the PLAAF to assess how those general principles might be implemented in specific potential combat operations; and, finally, use a combination of expert judgment and quantitative analysis to identify implications and responses for the USAF.

The previous Western analyses used for information on how the PLAAF intends to employ its forces in the event of a future conflict consisted largely of monographs and edited volumes published by RAND, the CNA Corporation, National Defense University, and the U.S. Army War College, along with articles published in journals, such as *International Security*, the Hoover Institution's *China Leadership Monitor*, and the Jamestown Foundation's *China Brief*. The Chinese military publications included books and journals published by organizations under the Chinese People's Liberation Army (PLA) [中国人民解放军] itself, as well as publications by closely affiliated entities, such as the Aviation Industry Press, which is controlled by China's state-owned aviation industry.

In assessing the information found in these Chinese military publications, particular weight was given to those sources that were reference works or textbooks. The authors of these publications are often committees or sometimes even identified as being an entire organization (e.g., the PLAAF or the Strategic Studies Branch of the Academy of Military Science [AMS]) and, as such, clearly represent official views of the PLAAF or the PLA. Lesser weight was given to monographs or articles written by individuals, not organizations or committees, who are often expressing their personal perspectives and thus do not necessarily speak for the entire PLAAF or PLA.

The most important single source for this study was the *China Air Force Encyclopedia*,[6] a two-volume, 1,400-page reference work comprising 4.3 million Chinese characters (equivalent to more than 2 million words) that contains information on topics ranging from air force military thought to air force systems engineering. It was published by the Aviation Industry Press in 2005 but clearly represents the official institutional position of the PLAAF. The director of the editorial committee was Qiao Qingchen, then the commander of the PLAAF, and the vice directors were Wang Chaoqun and He Weirong, deputy commanders of the PLAAF, and Liu Yazhou, a deputy political commissar of the PLAAF. The entry authors belong to various PLAAF organizations, units, and academies and schools.[7]

Other Chinese military publications included reference works and textbooks published by the PLA's principal publishing arm—Liberation Army Press—and by its National Defense University and AMS, along with monographs and articles published by the above three organizations as well as by the PLAAF and Aviation Industry Press. The most authoritative of these is *Study of Campaigns*,[8] published by the National Defense University Press in 2006, which provides a

6   People's Liberation Army Air Force (PLAAF) [中国人民解放军空军], 《中国空军百科全书》 [*China Air Force Encyclopedia*], Beijing: 航空工业出版社 [Aviation Industry Press], 2005.

7   PLAAF, 2005, preface.

8   Zhang Yuliang [张玉良], ed., 《战役学》 [*Study of Campaigns*], Beijing: 国防大学出版社 [National Defense University Press], 2006.

baseline for the principles, requirements, organization, and operational activities for different types of PLA campaigns. It includes roughly 50 pages of text on PLAAF campaigns, as well as sections on general campaign theory, joint campaigns, and the campaigns executed by other PLA services (including the Second Artillery, as well as the PLA Army and the PLA Navy [PLAN]). Comparison with a 2000 version of the *Study of Campaigns*[9] provided a basis on which to observe changes in Chinese campaign theory in the interval between the publication of the two documents.

Some authoritative Chinese source material was also available on PLAAF tactics, in the form of *Study of Air Force Tactics*,[10] a book produced by the PLAAF and published by the Liberation Army Press. Given this work's 1994 publication date, however, it was treated with caution and the information cross-checked against more-recent articles and reference works.

Two teaching texts from the National Defense University offer slightly less authoritative but more-detailed commentary on particular aspects of PLAAF campaign practice. The first of these is *Military Command Theory Study Guide*,[11] which treats both general topics associated with military command (e.g., command systems, the principles of command) and command practices for particular types of campaigns (e.g., "command in offensive air campaigns"). The second, *Campaign Theory Study Guide*,[12] covers more general subjects associated with military campaigns, as they are defined and practiced by the PLA. Perhaps because of their use as teaching materials, both of these sources are written in question-and-answer format and build from an explanation of basic terms and definitions to explanations of

---

[9] Wang Houqing [王厚卿] and Zhang Xingye [张兴业], eds., 《战役学》 [*Study of Campaigns*], Beijing: 国防大学出版社 [National Defense University Press], 2000.

[10] People's Liberation Army Air Force [中国人民解放军空军], 《空军战术学》 [*Study of Air Force Tactics*], Beijing: 解放军出版社 [Liberation Army Press], 1994.

[11] Lu Lihua [芦利华], 《军队指挥理论学习指南》 [*Military Command Theory Study Guide*], Beijing: 国防大学出版社 [National Defense University Press], 2004.

[12] Bi Xinglin [薛兴林], ed., 《战役理论学习指南》 [*Campaign Theory Study Guide*], Beijing: 国防大学出版社 [National Defense University Press], 2002.

more-complex operational practices and relationships. Because of the breadth of subjects covered, both are substantial, the former consisting of 639 and the latter 578 pages of dense text.

Finally, articles in the Chinese military press and nonauthoritative texts were also exploited. Given their status, these were treated with caution. Nevertheless, these sources often discuss individual topics in greater detail than the more-authoritative sources and are therefore sometimes useful for gaining an understanding of what Chinese operational activities might look like in practice. A single-authored text titled *Air Raids and Counter–Air Raids in the 21st Century*,[13] for example, describes in some detail what "air defense corridors" and the defense of cities and fixed points would entail, while *How Air Attacks and Air Defense Are Fought*[14] provides insight into the lessons learned by the PLAAF from the world's recent air campaigns. Moreover, while these sources may be somewhat less authoritative than the guides and textbooks mentioned earlier, some are clearly well researched and may be taken, at a minimum, as the views of well-informed sources. In the case of *Air Raids and Counter–Air Raids in the 21st Century*,[15] for example, authorship was provided by a committee of 17 authors led by a PLAAF major general at the National Defense University, while, in the case of *How Air Attacks and Air Defense Are Fought*,[16] authority is bolstered by the status of one of the coauthors as the lead author and editor of *Campaign Theory Study Guide*.[17]

It is important to point out that none of the primary sources used for this study should be considered *doctrine* or *doctrinal* as those terms

---

[13] Cui Changqi [崔长崎], Ji Rongren [纪荣仁], Min Zengfu [闵增富], Yuan Jingwei [袁静伟], Hu Siyuan [胡思远], Tian Tongshun [田同顺], Ruan Guangfeng [阮光峰], Hong Baocai [洪宝才], Meng Qingquan [孟庆全], Cao Xiumin [曹秀敏], Dai Jianjun [戴建军], Han Jibing [韩继兵], Wang Jicheng [王冀城], and Wang Xuejin [王学进], 《21世纪初空袭与反空袭》 [*Air Raids and Counter–Air Raids in the Early 21st Century*], Beijing: 解放军出版社 [Liberation Army Press], 2002.

[14] Peng Xiwen [彭希文] and Bi Xinglin [薛兴林], 《空袭与反空袭怎样打》 [*How Air Attack and Air Defense Are Fought*], 中国青年出版社 [Chinese Youth Press], 2001.

[15] Cui et al., 2002.

[16] Peng and Bi, 2001.

[17] Bi, 2002.

are used in the U.S. military context. The primary sources used for this study are not the equivalent of the U.S. Joint Publication (JP) series or Air Force Doctrine Documents (AFDDs).[18] Although, in many cases, they represent the official positions of the PLAAF or PLA, they are nonetheless reference works and textbooks, not official doctrinal publications that play the same role as the JP or AFDD series. Perhaps a closer equivalent of those documents in the PLA are its "guidance" [*gangyao* 纲要] and "combat regulation" [*zhandou tiaoling* 战斗条令] documents. For example, the PLA issues campaign guidance [战役纲要] documents for each of its services, including the PLAAF, as well as a joint campaign guidance document. According to the *China Air Force Encyclopedia*,[19] the PLAAF campaign guidance includes "standard military guidelines for PLAAF campaign operations" and is the "fundamental basis for the Air Force campaign group to organize campaign operations and exercises." Signed in 1999 by China's top military leadership, its contents include the following:

- the nature of air force campaigns, basic campaign types and campaign principles
- air force campaign organization for command and coordination mechanisms
- campaign guiding thought, operational duties, and operational methods for air force offensive campaigns, air defense campaigns, air blockade campaigns, and coordination with ground, naval, and Second Artillery Force campaign operations
- campaign electronic countermeasures, campaign airborne duties and demands
- demands and basic methods of campaign operational support: logistic support, armament support, and political support.

In addition to its overall campaign guidance, the PLAAF has combat regulations for "combined arms combat" [合同战斗条令] and

---

[18] U.S. JPs and AFDDs can be found at Joint Staff, "Joint Electronic Library," as of December 10, 2009, update.

[19] PLAAF, 2005, p. 328.

for fighter aviation, attack aviation, bomber aviation, reconnaissance aviation, transport aviation, SAM, AAA, airborne, electronic warfare (EW), radar, communications, chemical warfare defense, and "technical reconnaissance" force combat.[20] Campaign guidance and combat regulation documents are generally classified, however, and none of them was available for this study. The reference works and textbooks analyzed for this study are believed to be based on and consistent with these documents but should not be regarded as equivalent to them.[21]

It should also be noted that the primary sources used in this study, and probably the official campaign guidance and combat regulations as well, do not necessarily reflect actual current practice of the PLAAF or other parts of the PLA. In some cases, they refer to the employment of capabilities (e.g., low-observable aircraft) that the PLAAF does not yet possess, and, in other cases, resources or other limitations may also prevent the PLAAF from being able to operate its forces in the ways described in these sources. Rather than a depiction of how the PLAAF's forces are actually operated today, therefore, these publications appear to represent the views of the PLA, the PLAAF, and Chinese officers and analysts about how they *ought* to be employed. From the perspective of the present study, this is not necessarily a drawback, as these publications can thus be viewed as a description of how the PLAAF aspires to operate in the future, and the goal of the study is to identify the employment concepts the PLAAF might implement in a conflict occurring any time in the next decade or so, not just those that it would employ in a conflict that occurred in the near term. Indeed, given that the gap between PLAAF and U.S. capabilities is likely to shrink in coming years, understanding how the PLAAF would employ the capabilities it will have in the future is probably more important than understanding how it would employ its capabilities today. But it should also be acknowledged that, as a result of changes in thinking, technology, or assessments of the operating environment, future prac-

---

[20] PLAAF, 2005, pp. 328–330.

[21] The PLAAF also has a set of training and evaluation guidelines (*dagang* [大纲]) based, presumably, on the campaign guidance and combat regulations. See PLAAF, 2005, pp. 331–332, and Office of Naval Intelligence, 2007, pp. 28–29.

tices of the PLAAF may never precisely correspond to the employment concepts described in the Chinese military publications analyzed for this study.

The sources used for information about the posture and capabilities of the PLAAF include standard reference sources, such as the U.S. Department of Defense's annual report to Congress on Chinese military power,[22] various publications of Jane's Information Group, and certain reliable defense-related websites, such as Air Power Australia[23] and SinoDefence.com.[24]

Overview of This Monograph

This introductory chapter is followed by ten additional chapters. Chapter Two describes the organization of both the PLAAF and those elements of the PLA's other services that support or complement the missions of the PLAAF, including the PLAN's aviation forces, shore-based AAA, and surface naval forces (which operate SAM and AAA systems); the PLA Army's air defense (SAM and AAA) and aviation (helicopter) forces; and the Second Artillery Force's conventional surface-to-surface missiles (SSMs). Chapter Three describes how PLAAF employment concepts have evolved since the PLAAF's founding in 1949. Chapters Four through Nine describe Chinese concepts for the employment of air forces, based on the Chinese military publications described in this chapter. Chapter Four provides an overview of PLAAF employment concepts. Chapters Five through Eight provide detailed descriptions of the employment concepts of each of the four major types of air force campaign: air offensive campaigns, air defense campaigns, air blockade campaigns, and airborne campaigns, respectively. Chapter Nine describes the role of the PLA's other services in air force campaigns. Chapter Ten explores how the general principles described in Chapters Four through Nine might be operationalized in a specific poten-

---

[22] Office of the Secretary of Defense, 2008.

[23] "Air Power Australia," as of December 19, 2009, update.

[24] "SinoDefence.com," undated home page.

tial real-world campaign, and what the implications might be for U.S. force employment. Chapter Eleven discusses the implications of the findings of Chapters Four through Ten and provides recommendations for responding to those implications.

CHAPTER TWO

The Organization of China's Air and Missile Forces

To understand the organization of Chinese aerospace power, it is useful to examine how it fits into the overall structure of China's armed forces.[1] The Chinese leadership controls the Chinese military through the Central Military Commission (CMC), which, in recent years, has consisted of China's top civilian leader (currently Hu Jintao, secretary general of the Communist Party of China and president of the People's Republic of China, or PRC), and several top military leaders.[2] The CMC leads and directs China's military, called the PLA, through four general departments: the General Staff Department [总参谋部], the General Political Department [总政治部], the General Logistics Department [总后勤部], and the General Armaments Department [总装备部].[3]

---

[1] This chapter draws heavily from Allen, 2002, 2003; and Kenneth W. Allen, Glenn Krumel, and Jonathan D. Pollack, *China's Air Force Enters the 21st Century*, Santa Monica, Calif.: RAND Corporation, MR-580-AF, 1995.

[2] Hu Jintao serves as the chairman of the CMC. As of late 2010, the vice chairmen of the CMC were PRC Vice President Xi Jinping, Guo Boxiong, and Xu Caihou, and the other members of the CMC were Liang Guanglie (the Minister of Defense and previous Chief of the General Staff), Chen Bingde (the current Chief of the General Staff), Li Jinai (director of the General Political Department), Liao Xilong (director of the General Logistics Department), Chang Wanquan (director of the General Armaments Department), Jing Zhiyuan (commander of the Second Artillery), Wu Shengli (commander of the PLAN), and Xu Qiliang (commander of the PLAAF). See Cheng Li, "China's Midterm Jockeying: Gearing Up for 2012 (Part 3: Military Leaders)," *China Leadership Monitor*, No. 33, June 28, 2010, p. 2.

[3] Kenneth W. Allen and Maryanne Kivlehan-Wise, "Implementing PLA Second Artillery Doctrinal Reforms," in Mulvenon and Finkelstein, 2005, p. 168. Some analysts, including

The PLA consists of the PLA Army [陆军], the PLAN [中国人民解放军海军], the PLAAF [中国人民解放军空军], and the Second Artillery Force [第二炮兵], which controls all of China's land-based nuclear missiles and most of its conventionally armed SSMs. In principle, the PLAN and PLAAF are considered distinct "services" [军种], equal in status to the PLA Army (the Second Artillery is considered an "independent branch," not a full service like the PLAN and PLAAF, but is under the direct control of the CMC and, in many ways, operates as a separate service), but, in practice, the PLAN and PLAAF have traditionally been subordinate to the PLA Army. The PLA leadership has been dominated by officers from the PLA Army, and, prior to 2004, the PLAAF, PLAN, and Second Artillery commanders had an "army equivalent position" that was equivalent only to that of a military region (MR) commander (see below for a description of the MRs).[4] Due to the position of its commander within the PLA hierarchy, the PLAAF (along with the PLAN and Second Artillery) was "hindered in its ability to promote some programs and missions."[5]

This situation of inequality appears to have changed in recent years, however, as symbolized by the fact that, beginning in 2004, the commanders of the PLAN, PLAAF, and Second Artillery became members of the CMC.[6] Moreover, in July of that year, PLAAF Lieutenant General Xu Qiliang was promoted to Deputy Chief of General Staff of the PLA—the first PLAAF general to be appointed to this position since 1976.[7] In August 2003, PLAAF General Zheng Shenxia became the first air force general to head the AMS.[8] In August 2006,

---

Allen and Kivlehan-Wise, 2005, translate the Chinese name of this organization as General *Equipment* Department.

[4] Allen, 2002, p. 360. See also Allen, 2002, p. 360, fn. 851.

[5] Allen, 2002, p. 360.

[6] Kenneth W. Allen, "The PLA Air Force: 2006–2010," paper presented at the CAPS-RAND-CEIP International Conference on PLA Affairs, Taipei, November 10–12, 2005b, p. 2.

[7] "PRC General Xu Qiliang's Promotion to Spur Combined Operations of Armed Forces," *The Standard*, July 14, 2004.

[8] Allen, 2005b.

PLAAF Lieutenant General Ma Xiaotian was promoted to president of China's National Defense University.[9] And in December 2005, the deputy director position of the General Logistics Department was also filled by PLAAF Lieutenant General Li Maifu.[10]

For peacetime operations, China is divided up into seven MRs [军区]: the Shenyang, Beijing, Lanzhou, Nanjing, Guangzhou, Jinan, and Chengdu MRs (see Figure 2.1).[11] The commander of each MR—which, to date, has always been a PLA Army officer—has control over all PLA Army units, as well as all military operations, in his or her MR.[12] Similar to the U.S. armed forces, however, during peacetime, the Chinese navy, air force, and Second Artillery Force are responsible for operational command, training, and other administrative and management issues of their respective forces in each MR.[13] In the event of a war, a theater [战区] command would be established with operational command of all (conventional) military units within one or more MRs.[14]

China's aerospace power is distributed among all four major elements of the PLA: the PLA Army, the PLAN, the PLAAF, and the Second Artillery Force. The aerospace power elements of all of these

---

[9] "TKP: Ma Xiaotian to Become President of PLA National Defense University—Report by Wu Yue: 'Ma Xiaotian Promoted to Office of President of Defense University,'" *Ta Kung Pao*, August 18, 2006.

[10] Yang Xuejun [杨学军] and Zhang Wangxin [张望新], eds., 《优势来自空间：论空间战场与空间作战》 [*Advantage Comes from Space: On the Space Battlefield and Space Operations*], Beijing: 国防工业出版社 [National Defense Industry Press], 2006.

[11] Kenneth W. Allen, *People's Republic of China, People's Liberation Army Air Force*, Washington, D.C.: Defense Intelligence Agency, DIC-1300-445-91, April 15, 1991, pp. 10–13. Though dated, this is the seminal work on Chinese airpower.

[12] Dennis J. Blasko, "PLA Ground Forces: Moving Toward a Smaller, More Rapidly Deployable, Modern Combined Arms Force," in Mulvenon and Yang, 2002, pp. 309–345.

[13] Allen and Kivlehan-Wise, 2005, p. 168. Although the Second Artillery Force is controlled directly by the CMC, in wartime, conventional missile units would probably be put under the control of the theater commander.

[14] Allen and Kivlehan-Wise, 2005, p. 168.

Figure 2.1
China's Military Regions

SOURCE: Office of the Secretary of Defense, 2006, p. 25.
RAND MG915-2.1

organizations are described in the sections that follow, but this study focuses primarily on the PLAAF.

People's Liberation Army Air Force

The organization of the PLAAF is complex and includes an administrative structure, four operational branches, specialized support units, logistics and maintenance support units, academies and schools, and research institutes.[15] Its vertical chain of command consists of four operational and administrative levels (see Figure 2.2):

- *PLAAF headquarters.* Responsible for policy and training and equipping the air force, and oversees the Military Region Air Forces (MRAFs) as well as directly subordinate operational units and training and testing bases.[16]
- *MRAF* [军区空军] *headquarters.* Each MR has an associated MRAF. The MR commander is responsible for joint operations, while the MRAF commander (also a deputy MR commander) is responsible for air force operations within the MR.[17] Each MRAF oversees one or more command posts as well as directly subordinate operational units and training and testing bases.
- *Command posts* [空军指挥所]. Reporting to each MRAF headquarters are one or more command posts, which are responsible for a subset of operational units within the MRAF. (Operational units within an MRAF may report to a command post, be directly controlled by the MRAF headquarters, or be directly controlled by PLAAF headquarters.[18])
- *Operational units* [部队]. These may be directly subordinate to PLAAF headquarters, the MRAFs, or command posts. Their composition is described in the next sections.

Figure 2.2 illustrates the complete chain of command for the PLAAF.

The PLAAF's administrative structure parallels that of the PLA as a whole. Each administrative element, from air force headquarters

---

[15] Allen, 2002, p. 370.

[16] Allen, 2002, p. 382.

[17] Allen, 2002, p. 382.

[18] Information Office of the State Council of the People's Republic of China, *China's National Defense in 2006*, Beijing, December 2006; Allen, 2005b, pp. 5, 15, fns. 18–19; "China: Air Force," *Jane's World Air Forces*, February 23, 2007.

Figure 2.2
PLAAF Chain of Command

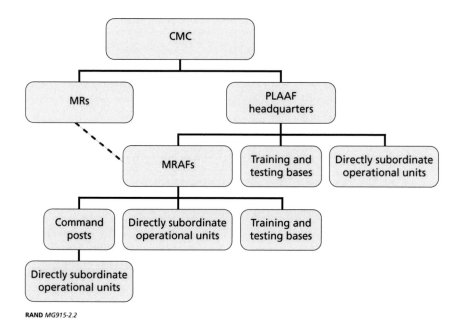

RAND *MG915-2.2*

to operational units, has a commander, political commissar,[19] two to four deputy commanders, and one or two deputy political commissars who oversee four departments—headquarters, political, logistics, and armaments—that mirror the four general departments of the PLA (see Figure 2.3).

The PLAAF consists of four combat branches [兵种]: aviation [航空兵], AAA [高射炮兵], SAMs [地空导弹兵], and airborne [空

---

[19] The political commissar system is a unique feature of PLA leadership. In principle, the commanders and the commissars share joint leadership at all levels of the Chinese military. While the commander takes the lead in military affairs, the commissar has responsibility for party organization and discipline, most aspects of personnel policy, and general unit morale and cohesion. The system has always been problematic, however, and, with the professionalization of the PLA, commanders have gained the status of first among equals within many Chinese military units and organizations. See You Ji, "Sorting Out the Myths About Political Commissars," in Nan Li, ed., *Chinese Civil-Military Relations: The Transformation of the People's Liberation Army*, New York: Routledge, 2006, pp. 89–116.

Figure 2.3
PLAAF Administrative Structure

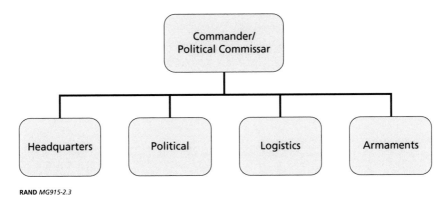

RAND *MG915-2.3*

降兵].[20] Note that, unlike in the USAF, the PLAAF's airborne forces include the actual foot soldiers, not just the aircraft used to transport them.

Often, a dichotomy is made between aviation and the other branches. Institutionally, they do not enjoy equal status. Air defense elements were merged with aviation in only 1957, and aviation still dominates in planning and decisionmaking.[21] Administratively, this historical legacy is reflected in the structure of the Headquarters Department of the PLAAF headquarters. Within the Headquarters Department, there are separate second-tier departments for radar, communications, and air defense (SAMs and AAA), but no second-tier department for the aviation branch. Instead, the Operations Department functions as the second-tier department for aviation.[22] This is because the entire organization was originally organized around aviation, with the other departments (e.g., the intelligence and training

---

[20] Information Office of the State Council of the People's Republic of China, *China's National Defense in 2002*, Beijing: New Star Publishers, 2002.

[21] See Allen, 2002, pp. 373–374.

[22] The Operations Department also functions as the second-tier department for airborne forces.

departments) supporting aviation.[23] In some respects, then, aviation enjoys the same status within the PLAAF that the army does within the PLA: Aviation is referred to as one of four branches, but it is the core of the entire service, with the other branches in support. As Allen observes, this situation may not be entirely rational, at least not based on historical performance. The PLAAF's combat record demonstrates that many more enemy aircraft have been destroyed by ground-based air defenses (AAA, mostly) than by Chinese fighters.[24]

PLAAF Aviation Unit Organization

The aviation branch is described as the PLAAF's "primary branch." It includes six varieties of aircraft: fighter aircraft (or J-class, according to the Chinese designators), ground attack aircraft (Q-class), bomber aircraft (H-class), fighter-bombers (JH-class), transport (Y-class), and reconnaissance aircraft (JZ-class).

From around 50 aviation divisions [师] during the early 1970s, the total number has dropped to fewer than 30 divisions, including 20 fighter, three bomber, three ground attack, and three transport.[25] Divisions typically have two or three regiments [团] of 24–40 combat aircraft each and, in some cases, an additional reconnaissance aircraft regiment.[26] Each aviation division has 72–120 fighters or 72–91 bombers.[27] Regiments serve as the basic tactical unit. Each regiment is further broken down into three flights [大队] of three sections [中队]

---

[23] Allen, 2002, pp. 404–408.

[24] Allen, 2002, p. 373.

[25] Office of the Secretary of Defense, *Annual Report to Congress: Military Power of the People's Republic of China 2009*, Washington D.C.: U.S. Department of Defense, 2009, p. 63, states that there are only two transport divisions. However, a third transport division was formed in the Chengdu MRAF in 2004–2005. Private communication with Kenneth Allen.

[26] Kenneth W. Allen, "Reforms in the PLA Air Force," *China Brief*, Vol. 5, No. 15, July 5, 2005a. Divisions and subordinate regiments are listed in "China: Air Force," 2007. This source suggests that the third regiment in some divisions may have been disbanded and that, where they exist, the three often share two airfields.

[27] "China: Air Force," 2007.

each.[28] In addition to the aviation divisions, there are several independent regiments (largely reconnaissance) and eight flight training academies (which, like divisions, each have three subordinate regiments).

PLAAF Ground-Based Air Defenses

The AAA and SAM branches, together with supporting radars and other systems, form the PLAAF's ground-based air defense system. Since the mid-1980s, many smaller AAA systems have been turned over to the army.[29] The general distribution of labor assigns 85mm and 100mm guns to the PLAAF, and smaller-caliber systems to the army (though the army also operates some of the larger-caliber systems). Since 1985, the PLAAF has disbanded most of its AAA divisions and SAM divisions, as well as some of its independent regiments, and created one composite (AAA and SAM) division [混成师].[30]

Currently, the only remaining air defense divisions—three SAM divisions and one composite AAA and SAM division—are in the Beijing-Tianjin area.[31] Each of these divisions likely has between two and four subordinate regiments. In addition, a number of independent SAM and AAA brigades and regiments survive. Each SAM brigade or regiment controls one to three battalions, and each AAA brigade or regiment has two to three battalions.[32]

---

[28] In English-language publications, the PLAAF and PLAN use *group* for the term we translate as *flight* [大队] and *squadron* for the term we translate as *section* [中队]. We have chosen to translate these terms as *flight* and *section*, as the U.S. usage of those terms better corresponds to the size of these units (eight to 14 aircraft for a flight and three to five aircraft for a section).

[29] Blasko, 2006, p. 135.

[30] Allen, 2002, p. 436.

[31] The estimate of unit numbers for divisions, regiments, and brigades is based on a variety of sources, which often differ substantially from one another. We have therefore made judgments about what the most-plausible numbers are, given equipment inventories and Chinese writings on specific units and air defense doctrine. Sources include International Institute for Strategic Studies, *The Military Balance 2007*, London: Routledge, 2007; "China: Air Force," 2007b; "People's Liberation Army Air Force," GlobalSecurity.org, as of April 27, 2005, modification; and Allen, 2002.

[32] Allen, 2002, pp. 435–436.

PLAAF Radar and Communications

At least one independent radar brigade and one independent regiment is assigned to each military region. Operationally, these are integrated into the MRAF command structure. Information is simultaneously passed to the PLAAF headquarters and through reporting chains within the MRAF. Each radar regiment has several battalions that collectively control as many as 25 radar stations [雷达站], which are company-level units. Each station operates two to three radars and is manned by around 20 officers and enlisted personnel.[33] The recent transformation of several regiments into brigade-sized elements suggests growth in the number of radar deployed. As of 2002, for example, there was no integrated, nationwide strategic air defense system, a situation frequently highlighted in PLA writing and one that PLAAF commanders and strategists wished to remedy. In late 2007, however, Xinhua News Agency reported that an "air intelligence radar network" [对空情报雷达网] covering the entire country had largely been completed, suggesting that the situation described in 2002 had been remedied.[34]

Communication units are also organized into regiments, each with about 1,600 personnel. One communication regiment is assigned to each MRAF headquarters.[35]

PLAAF Airborne Branch

Although the airborne forces may appear less relevant to air campaigns, Chinese employment concepts call for their integrated use against airfields and other relevant targets. Moreover, they are an integral part of the PLAAF and therefore more likely to actually be employed in this capacity than might be the case in other militaries. The 15th Airborne Army, roughly 35,000 strong, includes three airborne divisions: the

---

[33] Allen, 2002, pp. 436–437.

[34] See "National Air Intelligence Radar Network Realizes Complete Early Warning Coverage" 〈我国空情雷达网实现全域预警覆盖〉,《航空知识》 *Aerospace Knowledge*, No. 12, December 2007, p. 8.

[35] Allen, 2002, pp. 438–439.

43rd, 44th, and 45th. Each division is about 10,000 strong.[36] A special operations force (SOF) unit may be attached to the 43rd Airborne Division. Not surprisingly, the 13th Air Division (transport) is colocated with one of these units (the 45th) at Wuhan.[37] The 15th Airborne Corps also has its own subordinate transport regiment and helicopter flight.

Air and Missile Capabilities Controlled by Other Services

Aside from the PLAAF, several other services in the PLA operate aviation, SAM, or SSM forces.

Second Artillery

The Second Artillery Force was founded in 1966 under the direct control of the CMC and is headquartered in Beijing. Historically, the Second Artillery has been the least transparent part of the PLA. The Second Artillery's organizational structure, unique from those of the rest of the PLA services, is instructive and can provide insights into the Second Artillery's doctrine and strategy. The Second Artillery implements a "vertical command" [垂直指挥] system, which means that, unlike the PLAN and PLAAF, the MR headquarters normally have no command authority over Second Artillery units.[38] For example, unlike the MRAF and PLAN fleet commanders, the commander of a missile "base"[39] is not also a deputy commander of the MR in which it is located.

Like the PLA's other services, the Second Artillery has four departments: headquarters, political, logistics, and armaments. The

---

[36] The 43rd is based at Kaifeng, Henan; the 44th at Guangshui, Hubei; and the 45th at Wuhan, Hubei. The corps headquarters is located at Xiaogan, Hubei.

[37] "China: Air Force," 2007.

[38] Allen and Kivlehan-Wise, 2005, p. 168.

[39] The Second Artillery Force is subdivided into several subordinate "bases" (sometimes referred to as missile *armies*), each with several subordinate brigades. The "bases" are units of organization; the subordinate brigades are not literally colocated on a single base.

Second Artillery is estimated to have 100,000 personnel and includes an overall headquarters and six missile bases. Each base consists of several missile brigades, and each brigade consists of several launch battalions.[40] In addition, the Second Artillery comprises one engineering design facility, four research institutes, two command facilities, and possibly one early warning unit.[41] Although Second Artillery forces are not subordinate to MR commands during peacetime, during wartime, *conventional* missile forces may come under the control of the theater command (nuclear missile forces undoubtedly remain under the direct control of the CMC).[42]

People's Liberation Army Navy

The PLAN operates military aircraft, AAA, and shipboard air defenses, all of which might come into play in an air campaign. The PLAN's aviation forces possess aircraft and airfields; AAA, radar, communications, chemical defense, aircraft maintenance, and logistics units; and various academies.[43] The organizational and administrative structure of the PLAN's aviation forces has evolved over time, but, today, there are seven air divisions, which are assigned to 25 air bases located throughout three fleets.[44] Figure 2.4 illustrates the organizational structure of the PLAN's aviation forces.

According to *The Military Balance*, the PLAN operates roughly 800 fixed-wing combat aircraft, including 130 bombers, 350 fighters, and 300 fighter-bombers and attack aircraft. The PLAN has traditionally received older designs than the PLAAF, although China's most-capable fighter aircraft, the Su-30MK2, are operated by the PLAN. Until recently, PLAN aviation's force structure has been both antiquated and shrinking, and, given the ongoing retirement of older

---

[40] Allen and Kivlehan-Wise, 2005, pp. 169–170.

[41] Bates Gill, James Mulvenon, and Mark A. Stokes, "The Chinese Second Artillery Corps: Transition to Credible Deterrence," in Mulvenon and Yang, 2002, p. 521.

[42] See Gill, Mulvenon, and Stokes, 2002, pp. 527–528.

[43] Office of Naval Intelligence, 2007, p. 45.

[44] Office of Naval Intelligence, 2007, p. 46.

Figure 2.4
PLAN Aviation Force Organization

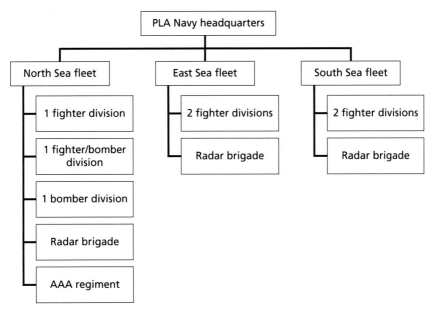

SOURCE: Office of Naval Intelligence, 2007, p. 47.
RAND *MG915-2.4*

models, it is unclear whether all of the aircraft listed in *The Military Balance* are still in the inventory. Numbers aside, however, the PLAN has, during the past several years, added new and more-capable classes of aircraft, including the Su-30MK2 and the JH-7 fighter-bomber.[45]

Air defenses for naval air bases and fleets are provided primarily by onboard ship systems and the PLAN's coastal defense AAA regiments. Onboard defenses have progressed rapidly in recent years, but modern air defense systems are still, nevertheless, only installed on the newest ships. Ship-mounted SAM defenses include medium- and high-altitude systems, such as the SA-N-7 (naval version of the SA-11) on the Hangzhou-class (Sovremennyy) and Guangzhou-class (Luyang I/Type 052B) destroyers, and the SA-N-6 (naval version of the S-300PMU/SA-20) on the Luzhou class, and the SA-N-6's indigenous

---

[45] IISS, 2007, p. 349.

equivalent, the HHQ-9, on the Luyang II class, as well as low- and very low-altitude systems, such as the HQ-61 and HQ-7 (Crotale) systems found on a wider variety of classes.[46]

PLA Army Air Defenses

PLA Army air defense elements are found in maneuver group armies, divisions, and regiments, as well as in separate coastal defense regiments and reserve air defense divisions. Within each of the 18 group armies, there is typically an AAA brigade or a composite SAM/AAA brigade, and some have more than one. In all, these group armies control 12 AAA brigades and nine composite SAM/AAA brigades. Each maneuver division also typically has an AAA regiment, and each brigade an AAA battalion. There are also eight coastal air defense regiments, 12 reserve AAA divisions, and two reserve AAA brigades.[47]

In total, we estimate that there are 25–27 SAM battalions, located primarily—if not exclusively—in the composite (SAM/AAA) brigades, and perhaps a total of 410–430 AAA battalions. The AAA battalions include 87 in the AAA brigades and composite air defense brigades of the group armies, 181 organic to maneuver divisions and brigades, and 142 in the reserve AAA divisions and coastal regiments.[48]

In an air campaign, the most relevant of these army air defenses would be coastal AAA regiments, reserve AAA divisions and brigades, and the air defenses of any group armies located in the combat zone. Coastal and reserve AAA units will be geographically arrayed, assigned to provincial commands that will then apportion resources to key targets, particularly cities, bases, power generation facilities, and so on.

PLA Army Aviation

The PLA's Army Aviation branch, which provides helicopter support to the PLA Army, was established in 1986, and the first units formed in 1988. Though estimates vary, the number of deployed helicopters

---

[46] IISS, 2007, p. 348.

[47] IISS, 2007, pp. 347–348; Blasko, 2006, pp. 39, 41.

[48] These estimates are based on unit compositions provided in Blasko, 2006, p. 42, cross-checked against equipment inventories found in IISS, 2007, pp. 347–348.

serving the PLA Army is unlikely to be much greater than 400. Most are transport or utility helicopters, with a smaller number of attack aircraft. This low number suggests that only a very few PLA Army elements have experience or training in air-mobile operations.[49] The PLA Army's deployed rotary-wing aircraft have been acquired from builders in China, Russia, the United States, and France. Organizationally, the PLA Army's aviation forces are grouped into nine regiments, one training unit, and four special aviation units subordinated to separate MR headquarters. Regiments are located within some, but not all, group armies.[50]

Air Force Campaign Command Arrangements

Air force campaigns will be waged either as PLAAF-only campaigns or, more often, as part of joint campaigns involving two or more services. In both cases, the PLAAF command structure at the operational level will include representatives from the various forces involved in the campaign, including from other services.[51]

Command Post Types

In any air force campaign, a basic command post and an alternative command post will usually be established. A forward command post and an airborne command post also may be established, depending on the situation.[52]

---

[49] Blasko, 2002, p. 323.

[50] Blasko, 2006, p. 43.

[51] Mark A. Stokes, "The Chinese Joint Aerospace Campaign: Strategy, Doctrine, and Force Modernization," in Mulvenon and Finkelstein, 2005, pp. 239–242. The Chinese word used to describe the operational level, *zhanyi* [战役] literally translates as *campaign*, and many analyses of Chinese military doctrine choose to use the English word *campaign* in these contexts. We have chosen to translate *zhanyi* as *operational* in this context because U.S. military writings normally use *operational* rather than *campaign* in these situations, and there appears to be no significant difference between this usage of *zhanyi* and the meaning of the U.S. term *operational*.

[52] Lu, 2004, p. 248.

Basic Command Post [空军战役基本指挥所]. A basic command post implements campaign commands for large air force formations. It is responsible for overall situational awareness and directing units' operations during a campaign. It will usually consist of the top military and political officials assigned to the campaign, a chief of staff, relevant headquarters personnel, and commanders from other supporting services and branches in charge of coordination and communication.[53]

Alternative Command Post [空军战役预备指挥所]. The alternative command post is established to replace the basic command post if needed. Usually it will consist of the campaign's deputy commanders and related command personnel. It can be deployed at the same time as the basic command post but will command only if the basic command post is unable to do so. It will receive the same information as the basic command post so that it can maintain situational awareness until it is activated in an emergency.[54]

Forward Command Post [空军战役前进指挥所]. The forward command post will direct air forces supporting ground or naval operations. It usually consists of deputy commanders and personnel under their leadership.[55]

Airborne Command Post [空军战役空中指挥所]. This is a mobile command element in a special air command or early warning aircraft. It is established to meet the needs of modern air combat, taking charge of operational theater command; it also can serve as a forward or alternative command post for ground operations. Chinese military sources state that one advantage of an airborne command post is that it is not easily attacked and factors on the ground do not influence its radar or communication equipment. In addition, it has a long range for surveillance and can detect aircraft or objects flying at a low altitude from a long distance. It consists of deputy commanders and their necessary staff.[56]

---

[53] Lu, 2004, p. 248.

[54] Lu, 2004, p. 248.

[55] Lu, 2004, p. 248.

[56] Lu, 2004, p. 248.

Command Types

Command can be centralized, dispersed, "normal" (hierarchical), and skip echelon.[57] In centralized command, the campaign commander makes most command decisions, which are delegated to him or her by authorities above the campaign commander. During the campaign, only the campaign commander can change deployments or make major decisions regarding the campaign. At the same time, according to Chinese writings, subordinates should handle certain problems according to "clear methods and requirements" to avoid overly centralized and rigid command.[58] They should propose changes to the original plan to the commander unless the situation has fundamentally changed and they are unable to contact their superiors—in which case they should take the initiative to handle the situation and report to the campaign commander later.[59]

In dispersed command, the campaign commander delegates command authority for most actions to subordinate commanders. The campaign commander still clarifies the operational missions, requirements, and forces necessary, but the specific methods and tactics are left up to the lower-level commanders. The campaign commander assigns weapons and forces to the unit carrying out the operations to maximize its independence, while also specifying the limits of the subordinates' powers and responsibilities.[60]

In normal (hierarchical) command, there is a clear division of labor for each command level, clear tasks for each unit, and clear relationships between subordinates and superiors—presumably planned in advance to form a "complete and systematic campaign command system."[61] According to Chinese texts, hierarchical command promotes more initiative and creativity and allows units to adjust to changing conditions. Chinese documents cite lack of efficiency as a drawback of

---

[57] Lu, 2004, pp. 249–250.

[58] Lu, 2004, p. 249.

[59] Lu, 2004, p. 249.

[60] Lu, 2004, p. 249.

[61] Lu, 2004, p. 250.

hierarchical command when there are several levels of command and extensive coordination.

In contrast with the multilayered command structure of normal command, in skip-echelon command, a campaign commander takes over command of operational units, skipping the commander below him or her. Skip-echelon command aims for maximal efficiency, especially in emergencies. The campaign commander is supposed to inform the "skipped" commander of any actions taken during skip-echelon command.[62]

Command and Coordination in the Campaign Planning Process

According to Chinese military texts, overall campaign plans are drawn up by the campaign command center, based on the coordinated instructions of the leadership and with input from representatives of services or branches involved in the campaign. The service branches draw up the plans for their own operations. All plans are supposed to include a worst-case scenario, contingency planning for the breakdown of automated systems, actionable details of forces used, and operational maneuvers. Plans are also expected to be flexible in the face of unanticipated events. Coordination is supposed to occur internally between aviation branches and between the services. Within the army, the forces that are most likely to be involved in coordination include army aviation, army artillery, and frontal offensive/defensive combat units. Within the navy, aircraft, naval surface ships, and naval surface-to-ship missile units are most likely to coordinate with the PLAAF on campaigns involving the PLAAF. The Second Artillery's "campaign tactical missile firepower units" (i.e., conventional SSMs) are most likely to coordinate with aviation-arm firepower units.[63]

Command Arrangements with Other Services and Branches During Joint Campaigns

PLA writings are vague with regard to coordination among services and branches during air campaigns. What is clear is that coordina-

---

[62] Lu, 2004, p. 249.

[63] Lu, 2004, pp. 254–255.

tion between various forces is carried out based on preestablished procedures rather than conducted in real time, which would require advanced command, control, communications, computers, intelligence, surveillance, and reconnaissance (C4ISR) systems. Coordination of missions, tactics, sorties, the area of operations, targets and their locations, and all of the forces involved (including cover and support forces and, as necessary, ground forces and plans for intercepts and escorts) is planned out in advance, as is the process for requesting additional cover and support during operations and for identifying friendly units.[64]

---

[64] Lu, 2004, pp. 254–255. See also Wang Houqing and Zhang Xingye, 2000, and Bi, 2002.

CHAPTER THREE

The Evolution of Chinese Air Force Doctrine

Fingerprints of core warfighting concepts that were advanced by Mao Zedong in the 1930s are still found in PLA and PLAAF employment concepts today. "Active defense," which is regarded as China's military strategy, was formulated by Mao as part of his "people's war" concept and is basically a strategy of weakness.[1] This strategy of weakness persists even to the present, since it is necessitated by the fact that potential PRC adversaries tend to have superior weapons and equipment. At its most basic level, active defense involves "taking tactically offensive action within a basically defensive strategy."[2] The parameters within which this strategy can be implemented are broad and can fall between the "active" end of the spectrum and the "passive," reactive end. The original goal of this strategy was to protect the PRC's large cities and industrial bases by using offensive operations to wear down an aggressor (in contrast to "passive defense").[3] As Chinese military capabilities have improved over time, however, the active defense strategy has evolved from stressing the "defense" aspect to stressing the "active" aspect in the form of a more offensively oriented strategy.[4]

---

[1] Blasko, 2006, pp. 95–96; Allen, Krumel, and Pollack, 1995, p. 23.

[2] Allen, Krumel, and Pollack, 1995, p. 24.

[3] Allen, Krumel, and Pollack, 1995, p. 25; John W. Lewis and Xue Litai, "China's Search for a Modern Air Force," *International Security*, Vol. 24, No. 1, Summer 1999, p. 66.

[4] Lewis and Xue, 1999, p. 81, provide the basis of this interpretation.

History

A major impetus for the PLA's emphasis on people's war and active defense strategies in the 1960s and 1970s was the belief among top PRC leaders not only that China's military was at a disadvantage in terms of weapons and equipment but that an invasion by the United States or Soviet Union was likely in the near term. Mao's fears of imminent global conflict in the 1960s and his advocacy of a strategy under which the tools of war would be manufactured in factories hidden in China's interior severely impeded the development of modern military capabilities.[5] Mao's perceptions also motivated violent political upheavals that occurred between 1958 and 1976. These events—the Great Leap Forward (1958), the Sino-Soviet split (1960), and the Cultural Revolution (1966–1976)—adversely affected the PLAAF's organizational and operational development.[6] After Mao's death in 1976, the net result of these political upheavals was a PLA that was weaker than it was in the 1950s, and one that stressed the defensive part of the active defense strategy more than the active part.

With the rise of Deng Xiaoping and attendant economic and political reforms in China in the late 1970s came a PLA strategy that was more attuned to Beijing's immediate military needs. Several milestones chart the PLA's important doctrinal evolution. In June 1985, the CMC declared that the likelihood of fighting a major, possibly nuclear, war was minimal and that China should instead concentrate its preparation on military conflicts along its periphery.[7] The shift in focus away from major conflict with great powers resulted in a rapid-reaction strategy based on the premises that China would be engaged only in local wars for the foreseeable future, that the PLA would need to strike to end the war quickly and meet political objectives, and that cost would be a big factor as equipment became more expensive to use

---

[5] Lewis and Xue, 1999, p. 67.

[6] Allen, Krumel, and Pollack, 1995, p. xvii.

[7] Allen, Krumel, and Pollack, 1995, p. 29.

and replace.[8] Following the 1985 shift in strategy, Chinese military journals indicated five types of wars on which the PLA should focus:

1) small-scale conflicts restricted to contested border territory, 2) conflicts over territorial seas and islands, 3) surprise air attacks, 4) defense against deliberately limited attacks into Chinese territory, and 5) "punitive counterattacks" launched by China into enemy territory to "oppose invasion, protect sovereignty, or to uphold justice and dispel threats."[9]

The last of these is an obvious reference to China's incursion into Vietnam in 1979, and, as suggested by the second item, Britain's 1982 conflict with Argentina over the Falkland Islands undoubtedly also influenced the Chinese conception of likely future wars.

The 1991 Persian Gulf War sent shockwaves throughout China's military community and accelerated the PLA's modernization and shifts in strategy. The United States' overwhelming dominance in that conflict led Chinese military leaders to push for advanced military technologies. According to Allen, Krumel, and Pollack, China's National Defense University recommended that the PLA

1) reduce the number of soldiers and improve the armed forces' equipment, training quality, and actual combat capability; 2) give priority to conventional arms over nuclear weapons; 3) introduce high-technology, including advanced guidance systems, pinpoint accuracy bombing, weapons of mass destruction, and stealth aircraft; and 4) build a rapid-response force.[10]

Chinese military writings began stating that the PLA must be capable of winning "local wars under high-technology conditions" [高技术条件下局部战争]. In China's 2004 national defense white paper,

---

8   Allen, 1997, p. 223.

9   Allen, Krumel, and Pollack, 1995, p. 29.

10   Allen, Krumel, and Pollack, 1995, p. 33.

this description was reformulated to "informationalized local wars" [信息化局部战争].[11]

This focus on high-technology warfare particularly emphasized airpower. Given that U.S. success in the Persian Gulf War was due in large part to overwhelming domination of the air, senior PLA leaders began to appreciate the implications of superior airpower.[12] The 1996 Taiwan Strait crisis and 1999 NATO operations over Kosovo further reinforced this appreciation, and China continues to digest the lessons learned from U.S. operations in Iraq and Afghanistan.[13]

Doctrine

Like the U.S. Air Force, the PLAAF was founded as part of China's army. However, unlike the USAF, which has developed employment concepts and doctrine independent of the U.S. Army's, the PLAAF's doctrine, despite progression since 1949, has struggled to move out of the army's shadow.[14] PLAAF doctrine has mostly evolved in step with that of the PLA ground forces. While the PLAAF was formally established on November 11, 1949, during these early years, "no consideration was ever given to making the air force a service independent of the army . . . because the PLA leadership did not want an autonomous aviation force."[15] Accordingly, the PLAAF's first commander and political commissar were chosen directly from the army.[16] The shadow cast

---

[11] Information Office of the State Council of the People's Republic of China, *China's National Defense in 2004*, Beijing, December 27, 2004. Subsequently, the English neologism *informationalized* has been reduced to *informationized*. See Information Office of the State Council of the People's Republic of China, 2006.

[12] Allen, Krumel, and Pollack, 1995, p. 32.

[13] Office of the Secretary of Defense, *Annual Report to Congress: Military Power of the People's Republic of China 2006*, Washington D.C.: U.S. Department of Defense, 2006, p. 5.

[14] Allen, 2002, p. 364. As discussed shortly, there are more-recent indications that the PLAAF is making headway toward becoming more of an independent service.

[15] Allen, Krumel, and Pollack, 1995, p. 37.

[16] Allen, Krumel, and Pollack, 1995, p. 35.

by the PLA over the PLAAF is evident in the early roles and missions of the Chinese air force. For example, the PLAAF's first operational mission in 1949—defending Beijing and Shanghai against Nationalist air raids—was defensive in nature.[17] Beginning in the early 1950s, one of the PLAAF's primary goals was to seize air superiority [夺取制空权] over the battlefield.[18]

The Korean War, battles over Taiwan's offshore islands, and the Vietnam conflict shaped the evolution of China's air force employment concepts, and the tempo of air and space power growth. During the Korean War, the PLAAF's original air plan was to support ground troops as its primary mission, a reflection of the PLA Army's influence on Chinese air strategy.[19] The PLAAF was unable to execute this strategy because of various technical limitations and had to change its mission to that of conducting air operations against U.S. forces. This, in turn, helped the PLAAF develop basic air defense strategy and tactics.[20]

Air operations against Nationalist forces on Taiwan's outlying islands of Yijiangshan and Jinmen (the latter also known as Quemoy or Kinmen) in the late 1950s also helped to shape Chinese air force employment concepts. The Yijiangshan Island campaign of 1954–1955 is the only campaign in PLA history to have involved combined air, ground, and naval operations.[21] The PLAAF's goals were to achieve air superiority, attack Taiwanese resupply ships, conduct decoy and reconnaissance missions, and provide direct air support for landing operations.[22] Lessons learned from the Yijiangshan Island campaign were to resonate in subsequent PLAAF strategy and employment concepts and

---

[17] Allen, Krumel, and Pollack, 1995, p. 101.

[18] Allen, 2002, p. 370.

[19] Zhang Xiaoming, "Air Combat for the People's Republic: The People's Liberation Army Air Force in Action, 1949–1969," in Mark A. Ryan, David Michael Finkelstein, and Michael McDevitt, eds., *Chinese Warfighting: The PLA Experience Since 1949*, Armonk, N.Y.: M. E. Sharpe, 2003, pp. 271–272.

[20] Zhang Xiaoming, 2003, pp. 271–272.

[21] Zhang Xiaoming, 2003, p. 279.

[22] Zhang Xiaoming, 2003, p. 280.

include a "relentless use of an overwhelming striking force to attack enemy artillery and firepower positions as well as command and communication centers."[23] Chinese military leaders also learned that they could overcome the short ranges and limited loiter times of their fighter jets by using the numerical superiority of PLAAF fighters to maintain continuous fighter patrols.[24] The third lesson was that, while attack sorties should be flown according to plan, commanders should allow flexibility "in target selection based on the need of ground forces."[25] In sum, the Yijiangshan experience reflected the PRC's concept of airpower's role in a local conflict.[26] In terms of PLAAF campaign theory, emphasis was placed on "air defense first, followed by air superiority, and then offensive air support."[27]

The Jinmen campaign of 1958, the most recent Chinese military conflict to truly involve air combat, was also an important shaper of PLAAF strategy and employment concepts. Among others, the conflict provides an example of how air operational principles were governed by rules from the very top—the CMC.[28] According to Zhang Xiaoming, these operational principles of the CMC stressed

> (1) using overwhelming force to achieve protection of forces and destruction of enemy forces; (2) subservience of military battles to political battles by a strict adherence to CMC operational policy; and (3) study and application of PLAAF experiences and tactics drawn from the Korean War.[29]

Because the PRC leadership was uncertain about the PLAAF's counterstrike capabilities vis-à-vis Taiwan, PLAAF doctrine remained defensive. Thus, it "deployed large numbers of fighters to the region

---

[23] Zhang Xiaoming, 2003, p. 282.

[24] Zhang Xiaoming, 2003, p. 282.

[25] Zhang Xiaoming, 2003, p. 282.

[26] Zhang Xiaoming, 2003, p. 282.

[27] Zhang Xiaoming, 2003, p. 282.

[28] Zhang Xiaoming, 2003, p. 283.

[29] Zhang Xiaoming, 2003, p. 284.

but could not capitalize on its numerical superiority," since it had to retain half of its aircraft to protect home bases.[30] In addition to political concerns of not wanting to escalate the Jinmen campaign into an international crisis, the limited range of Chinese MiG-17 aircraft also limited the operational capabilities of the PLAAF.[31]

Aside from battle experience as a determinant and molder of strategy and doctrine, political upheavals in the communist regime also had profound effects on the evolution of Chinese air force doctrine. Beginning with the Sino-Soviet split in the 1960 and followed by the Cultural Revolution, which festered until 1976, Chinese airpower, and the ability to execute its strategy and doctrine, atrophied. The Sino-Soviet split's primary effect on the PLAAF was to significantly slow modernization efforts, as China was highly dependent on Soviet technology transfers for equipping the PLAAF.[32] And, due to the fact that an air force is, by its very nature, a more technically oriented service than the army, the PLAAF suffered greatly during the Cultural Revolution, which eschewed anything having to do with intellectualism and expertise. Furthermore, the PLAAF's association with Defense Minister Lin Biao's failed coup attempt against Mao in 1971 resulted in it being marginalized until after Mao's death and the rehabilitation of Deng Xiaoping in 1978.[33] Partly as a consequence, PLAAF involvement during China's war with Vietnam in 1979 was limited. As in the case of the Jinmen conflict, China's air involvement during the conflict was constrained both by political factors—not wanting to involve the United States in the former case and the Soviet Union in the latter—and by the limited capabilities of the PLAAF.

Deng ushered in a new era of economic and military reform, which set all military services on a path to modernization and reform. Indeed, after Deng took control of the CMC and later became China's undisputed leader in 1978, he "elevated his perspective on airpower

---

[30] Zhang Xiaoming, 2003, p. 288.

[31] Zhang Xiaoming, 2003, p. 288.

[32] Allen, Krumel, and Pollack, 1995, p. 71.

[33] Allen, Krumel, and Pollack, 1995, p. 73.

to official CMC dogma."[34] This perspective viewed the pursuit of air superiority as crucial to Chinese military power and winning future wars.[35]

The actual implementation of Deng's directives on Chinese airpower modernization, however, was constrained during most of his tenure as China's paramount leader, for two major reasons. First, by attaching special political weight to the PLAAF, Deng not only wanted to alleviate the decrepit state of Chinese airpower; he also wanted to keep tight control over the PLAAF so as to prevent it from becoming the politically dangerous service it had been under Lin Biao during the Cultural Revolution.[36] Second, the army-centric mentality ingrained during the Mao era attenuated efforts to implement near-term improvements in the PLAAF.[37] For example, when the PLA began reorganizing ground forces into group armies in the early 1980s, the PLAAF was given guidance that its role was to support the needs of ground forces and that a victory was a ground force victory.[38]

The Gulf War of 1991 spurred renewed debate within the PLAAF and Chinese military establishment about how to modernize and develop Chinese airpower. The U.S. show of force in the Taiwan Strait crisis of 1996, in which the United States deployed two aircraft-carrier battle groups near Taiwan in response to Chinese military intimidation of Taiwan, further motivated doctrinal reform and technological modernization efforts in the PLAAF. The PLAAF's desire for a strategy of "quick reaction," "integrated coordination," and "combat in depth" had to be operationalized.[39] *Quick reaction* meant launching an instan-

---

[34] Lewis and Xue, 1999, p. 70.

[35] Lewis and Xue, 1999, p. 70.

[36] Lewis and Xue, 1999, pp. 70–71. Because of these constraints, the PLAAF remained subservient to the PLA's and other strategic priorities.

[37] Lewis and Xue, 1999, p. 74. The Mao-era dogma of self-reliance was relaxed to permit acquisition of foreign air-launched weapons and avionics. Only the purchase of foreign aircraft remained prohibited.

[38] Lanzit and Allen, 2007, pp. 439–440.

[39] Lewis and Xue, 1999, p. 79.

taneous retaliatory strike for deterrence, or even survival.[40] *Integrated coordination* meant allowing the air force to "manage the long-range bomber air groups and oversee the initial stages of joint operations with the other services and between air combat units stationed in different military regions."[41] Finally, *combat in depth* meant conducting operations over a wide geographical area.[42] However, operationalizing these concepts was difficult because, during the early 1990s, military reform tended to stress internal organization and structural changes, as opposed to training and equipment modernization.[43] The PLAAF lacked the equipment and training needed to implement this strategy.[44]

In the 1990s, PLAAF employment concepts assumed that future wars would be conducted according to an active defense strategy with three phases: "strategic defense, strategic stalemate, and strategic counterattack.".[45] Still under the umbrella of active defense, PLAAF campaigns were divided into two categories—defensive campaigns and attack campaigns—either of which could be one of two types: independent air force campaigns, and air force campaigns part of a joint campaign.[46] PLAAF publications also specified three levels of scale for an air defense campaign, with small campaigns requiring air defense of a strategic position, large campaigns requiring air defense of a battle area, and larger campaigns requiring air defense of many battle areas.[47]

A PLAAF study published in 1990 revealed both the desire to have a more unified air strategy, and the gap between desired strategy and the ability to implement it. For example, one challenge to execution of the aforementioned rapid-reaction strategy was the lack of a

---

[40] Lewis and Xue, 1999, p. 80.

[41] Lewis and Xue, 1999, p. 80.

[42] Lewis and Xue, 1999, p. 80.

[43] Allen, Krumel, and Pollack, 1995, p. 105.

[44] Allen, Krumel, and Pollack, 1995, p. 109.

[45] Allen, Krumel, and Pollack, 1995, p. 111. These phases are clearly based on Mao's writings and the PLA's experience in the Chinese civil war.

[46] Allen, Krumel, and Pollack, 1995, pp. 111–112.

[47] Allen, Krumel, and Pollack, 1995, p. 112.

unified air defense plan in the PRC.[48] Since each service possessed its own air defense forces, and coordinating the different elements *within* each service was challenging enough, it was virtually impossible to coordinate operations across services.[49]

Other dimensions of the PLAAF's strategy included two principles: "light front, heavy rear" [前轻后重] and a "deploying in three rings" concept.[50] The "light front, heavy rear" principle stemmed from the PLAAF's responsibility to protect airfields, "national political and economic centers, heavy troop concentrations, important military facilities, and transportation systems," and resulted in most fighter airfields, and almost all SAMs, being concentrated around China's large cities—most of which are at least 200 km from China's nearest borders.[51] Under "light front, heavy rear," the PLAAF "would organize its SAM and AAA forces into a combined high, medium and low altitude and a far, medium and short distance air defense net."[52] Intercept lines and aviation forces would also be organized into a series of interception layers.[53] However, in executing this concept, the PLAAF faced two daunting challenges: the limited range of PRC aircraft, and adversaries that had aircraft capable of conducting deep strikes into Chinese territory.[54] The limited range of PLAAF aircraft was worsened by the fact that most airfields and almost all SAMs were concentrated near China's large cities, far away from China's borders.[55] For the "light front, heavy rear" principle to work, moreover, the PLAAF needed to develop a better command-and-control system; otherwise, there was a risk of fratricide to friendly aircraft from SAMs and AAA.[56] Finally, because

---

[48] Allen, Krumel, and Pollack, 1995, p. 113.

[49] Allen, Krumel, and Pollack, 1995, p. 113.

[50] Allen, Krumel, and Pollack, 1995, pp. 114–115.

[51] Allen, Krumel, and Pollack, 1995, p. 114.

[52] Allen, Krumel, and Pollack, 1995, p. 114.

[53] Allen, Krumel, and Pollack, 1995, p. 114.

[54] Allen, Krumel, and Pollack, 1995, p. 115.

[55] Allen, Krumel, and Pollack, 1995, p. 114.

[56] Allen, Krumel, and Pollack, 1995, pp. 114, 116, 124.

of equipment and command-and-control limitations, the most challenging problem for PLAAF was the task of ground-force support.[57]

To be used in conjunction with the "light front, heavy rear" principle, "deploying in three rings" involved organizing a small quantity of interceptors, AAA and SAMs "as a combined air defense force into 'three dimensional, in-depth, overlapping' firepower rings."[58] Furthermore, according to Allen, Krumel, and Pollack,

> Each weapon system would be assigned a specific airspace to defend—high, medium or low. In-depth rings means assigning each weapon system a specific distance from the target to defend—distant, medium or close. Overlapping rings means organizing each weapon system into left, middle or right firepower rings facing the most likely avenue of approach.[59]

In 1993, after the Gulf War, 60 airpower specialists formed an airpower theory, strategy, and development study group to investigate independent air campaigns.[60] According to one study, by 1997, the Chinese air force had "claimed precedence over the other service branches, and the People's War as a unifying dogma had given way to service-specific strategies."[61]

According to another study, as of the late 1990s, the primary PLAAF missions were air coercion, air offensives, air blockades, and support for ground force operations.[62] Coercion could come in the

---

[57] Allen, Krumel, and Pollack, 1995, p. 118.

[58] Allen, Krumel, and Pollack, 1995, p. 115.

[59] Allen, Krumel, and Pollack, 1995, pp. 115–116.

[60] Stokes, 2005, p. 246.

[61] Lewis and Xue, 1999, pp. 89–90.

[62] Stokes, 2005, p. 247. The Chinese term for coercion, *weishe* [威慑], is translated by many analysts, including Stokes, 2005, as *deterrence*. As Stokes himself argues, however, *weishe* actually encompasses both deterrence, as it is normally understood, and *compellence*—forcing an adversary to do something it would not otherwise wish to do. The more accurate translation of *weishe*, therefore, is *coercion*, which, in Western strategic writings, also includes both deterrence and compellence. Stokes, 2005, p. 247, also uses the term *air strikes* rather than *air offensives*. In other parts of this monograph, however, we translate the Chinese term

form of demonstrations, such as deployments and exercises, weapon tests, or overflights. It could also come in the form of limited strikes to warn or punish an adversary. Air offensives, by contrast, would entail large-scale strikes with the goal of rapidly gaining air superiority, reducing an adversary's capacity for military operations, and establishing the conditions necessary for victory. An air blockade would entail attacks on airfields and seaports as well as air, land, and sea transportation routes with the goal of cutting an enemy off from contact with the outside world. Support for ground force operations would include attacks on logistics facilities, hardened coastal defenses (in the case of an amphibious operation), reinforcements, and key choke points, such as bridges. It would also include battlefield close air support, strategic and theater airlift, airborne operations against command headquarters, and the deployment of ground-based air defenses to protect ground forces and key facilities.[63]

According to Stokes, as of the late 1990s, PLAAF operational principles included "surprise and first strikes," "concentration of best assets," "offensive action as a component of air defense," and "close coordination." *Surprise and first strikes* refers to the goal of crippling an opponent and gaining the initiative early in a conflict through surprise and large-scale attacks on key targets, such as the enemy's air command-and-control structure, key air bases, and SAM sites. *Concentration of best assets* supports this principle and refers to using the PLAAF's best assets in the initial strikes and to dedicating the majority of them to targets that will have the most influence on a campaign. *Offensive action as a component of air defense* refers to using offensive counter–air attacks as an integral aspect of air defense by attacking those enemy assets that pose the greatest threat. *Close coordination* refers to coordinating the air assets of all services (Army, PLAN, PLAAF,

---

to which he is referring, *kongzhong jingong* [空中进攻], as *air offensives*. (We translate other Chinese terms, such as *kongzhong tuji* [空中突击] or *kongzhong daji* [空中打击], as *air strikes*.) For consistency and accuracy, therefore, we use *air offensives* here instead of Stokes's *air strikes*.

[63] Stokes, 2005, pp. 247–250.

Second Artillery), as well as unified command at the theater level.[64] As seen in Chapters Four through Eight, these principles remain key elements of PLAAF employment concepts.

A major change in PLAAF doctrine occurred in 1999, when it revised its campaign guidance [纲要], which "provides the classified doctrinal basis and general guidance for how the PLAAF will fight future campaigns."[65] Since the guidance is classified, its exact contents are unknown. What Western analysts do know is that the guidance shows that the PLAAF had deepened its understanding of the operational level of war. The PLAAF was also now tasked with preparing for three types of air force campaigns: air offensive, air defense, and air blockade.[66]

Until 2004, the PLAAF lacked its own, service-specific strategy, and the actual ability of the PLAAF to integrate its campaign and operational principles with the Second Artillery, PLA Army, and PLAN was questionable. One study states that, until that time, the Chinese air force relied "almost solely on the PLA Army's 'Active Defense' operational component as its strategic-level doctrinal guidance."[67] The approval of the PLAAF's active defense strategy as a component of the National Military Strategic Guidelines for air operations in 2004, however, indicated an important shift in the PLAAF's status.[68] The PLAAF's strategic component of the National Military Strategic Guidelines is now identified as "'Integrated Air and Space, Simultaneous Offensive and

---

[64] Stokes, 2005, pp. 250–254.

[65] Allen, 2005b, p. 3.

[66] Allen, 2005b, p. 4. These were three of 22 different types of campaigns identified by the PLA. The remaining 19 campaigns include five ground force campaigns (mobile warfare, positional offensive, urban offensive, positional defensive, and urban defensive); six naval force campaigns (sea blockade, sea lines of communication destruction, coastal raid, antiship, sea lines of communication defense, and naval base defense); two Second Artillery campaigns (nuclear counterattack and conventional missile campaigns); and six joint service campaigns (blockade, landing, anti–air raid, border counterattack, airborne, and antilanding).

[67] Lanzit and Allen, 2007, p. 448.

[68] Lanzit and Allen, 2007, pp. 450–451.

Defensive Operations' [空天一体，攻防兼备]."[69] While it does not appear that the PLAAF yet has a service-specific strategy that is as well defined as that of the PLAN—that of offshore defense—it does seem that the PLAAF is now seen as a truly independent service. The same study cites Hong Kong press reports stating that the PLAAF should be a strategic air force that stands "side by side" with the Chinese army and navy "to achieve command of the air, ground, and sea."[70]

---

[69] Yao, 2005, p. 57, quoted in Lanzit and Allen, 2007, p. 450.

[70] Lanzit and Allen, 2007, p. 451.

CHAPTER FOUR

Chinese Concepts for the Employment of Air Forces

Chinese concepts for the employment of air forces are developing quickly. PLAAF texts in the early 1990s noted that, in modern combat, air operations take place over a larger space than in the past, the battlefield situation can change rapidly, EW is more intense, PGMs are used more frequently, services' operations are more integrated, and combat is more destructive.[1] More recently, areas of airpower that have been described as receiving increased attention include expanding the functions and missions of airpower, improving "comprehensive warfighting capabilities" [综合作战能力], new technologies to enable these capabilities (such as high-technology aircraft, missiles, and artillery, as well as C4ISR and electronic countermeasures, information, and technologies for integrating air and space), and personnel and training.[2] As airpower continues to change, Chinese military writings note that integration is critical: integration of different kinds of air platforms (such as tankers, transport aircraft, fighters, and bomber aircraft), integration of air and space, and integration of ground, naval, army, space, and electromagnetic power.[3]

A key theme that appears throughout PLA texts is that the operational space of air combat is becoming deeper and more three-

---

[1]   See, for example, PLAAF, 1994, p. 93.

[2]   PLAAF, 2005, p. 55.

[3]   "Combined" [总体] and "integrated" [一体化] are frequent themes throughout PLAAF writings, especially in the context of the integration of air and space and integration of other services and branches with airpower.

dimensional, and increasingly incorporates stealthy operations. Overwhelming force is used, and it is used preemptively, continuously throughout the conflict, and carefully in a way that will attack and destroy vital targets while reducing collateral damage. Operations take place at all times of day and in all weather. Different types of engagements are said to be developing: "Noncontact" (i.e., long-range, standoff), integrated air and space, and integrated information and firepower operations are all examples.

Finally, PLA publications note that it is increasingly difficult to penetrate enemy defensive lines due to developments in early warning, EW, and high-performance fighters. PLA writings indicate that these are characteristics of modern air combat that successful air forces need to possess or be prepared to counter. According to Chinese military writings, the evolution in airpower theories is ongoing: Local wars experiment with new ideas, and, with each new conflict, air force theories change.[4]

Chinese military publications also note that there have been several changes in the operational methods employed by the air forces of leading airpower states. Specific areas of significant change in recent years include integrated air offensive operations, integrated defensive operations, striking over the entire depth of the battlefield (at all distances and at all altitudes), using elite forces at the beginning of the war to move first against the enemy, over-the-horizon combat, long-range standoff munitions [远距离投射弹药] for ground attack, precision strike to destroy key points, stealthy strikes, superbombers, continuous offensive operations in all weather and at night, and unified command and concentrated control.[5] Not surprisingly, most of these seem to be areas in which the PLA is trying to improve, both conceptually in its official writings and operationally in its training and exercises. All of these concepts are themes that appear repeatedly in military writings on the changes in airpower, how to counter enemy strikes, and how air operations are conducted.

---

[4]   PLAAF, 2005, p. 55.

[5]   PLAAF, 2005, pp. 82–83.

While some of the employment concepts described in this monograph are aspirational, the concepts described probably will, for this very reason, "age" well as the PLA develops improved operational capabilities. For example, the PLAAF has had employment concepts for operating at night, in all weather, and with advanced weapons and fighters, since the mid-1990s, even though the PLAAF did not then operate well or consistently at night or in all weather conditions. Having employment concepts in place already will likely assist the transition from aspiration to actual capability.

The PLA clearly believes that having air, information, and space superiority is vital to winning campaigns and will be even more vital in future wars. In recognizing the PLA's limitations, however, military texts limit the need for gaining air, information, and space superiority to a "certain time and space" necessary to satisfy tactical, operational, or strategic objectives. Air, information, and space superiority is viewed as a means for achieving campaign (or strategic) objectives—not as an end in itself. Military writings distinguish between tactical, operational, and strategic dominance and do not assume that air, information, or space superiority must be strategic or absolute.[6]

Related to the importance of having air, information, and space superiority is the perceived need to improve command and control, which is viewed as increasingly vital for successful campaigns. PLA writings still stress the importance of the commander or command element as the key decisionmaker and actor in campaigns. The command organization could be in an aircraft (perhaps someday an AWACS aircraft) or on the ground, but, for the most part, the commander—not the pilot—makes most decisions. There are exceptions for times when the pilot has information that the command center does not, especially if there is electronic interference, but, overall, the stress still is on the commander's judgment and on retaining electromagnetic superiority so that this situation never arises.[7]

Joint operations are perceived as increasingly important, and the air campaigns discussed in this monograph will generally be part of a

---

[6] PLAAF, 2005, pp. 39, 48.

[7] PLAAF, 2005, pp. 156–166.

joint campaign (although they can also be conducted as independent campaigns). However, joint operations for the PLA are still limited by its weakness in performing joint exercises and by technical and other problems in command-and-control capabilities. These limitations are not explicitly referenced in military writings about employment concepts, but the lack of joint interoperability is nevertheless implied. Except for mention of individual service representatives in command organizations and of certain missions that would be performed by different services (such as Second Artillery strikes on targets in the enemy's rear areas and naval strikes against targets close to the shore), there is little discussion of the different roles that each of the services could play in a campaign and how they would coordinate their roles.

A final question is whether the PLAAF will become a more "strategic" air force, capable of achieving—or at least aspiring to achieve— political objectives through strategic air strikes. The PLAAF clearly aspires to become a more offensive air force that takes the initiative, uses preemptive strike and surprise, and carries out strikes against "key targets" with PGMs, all of which could presage an air force capable of achieving strategic objectives. At the same time, the PLAAF remains mostly a defensive force, and a lot of its responsibilities remain tied to supporting the other services. The PLAAF's tactics still emphasize defense: Positioning, camouflage, and ambush all remain essential elements of PLAAF tactics. While none of these tactics rules out becoming a "strategic" air force, taken together, they imply that the PLAAF still expects to face a superior enemy and to act defensively as often as it acts offensively.

PLAAF Strategy

Two critically important concepts that come up repeatedly in writings on air force employment concepts are the integration of air and space and preparing both the offensive and defensive. According to the *China Air Force Encyclopedia*, these two concepts have been at the center of air force strategy since 2004: "In 2004, the Central Military Commission . . . established the PLAAF strategy of 'integrated air and space, and preparing simultaneously for the offensive and the defensive.'"[8] Their identification as the essence of air force strategy reflects a significant shift as the PLAAF has moved toward trying to build a military that will integrate space-based information and operations and a more offensive orientation.

Integrated Air and Space

While integrated air and space operations is a central theme and apparently an aspiration for the PLAAF, it seems to be a concept that still is not very well developed. For example, while official sources on air and space integration readily note that other major air forces have already integrated air and space (notably the United States and the former Soviet Union) and even go so far as to state that "following the development of space technology's speed of flight and use in actual combat, strategic air strike will develop into strategic air and space strike,"[9] no mention is made of how the PLAAF intends to achieve the "strategy" of air and space integration. This is the case even though the *China Air Force Encyclopedia* notes that control of space will have a great impact on "future comprehensive warfare" and refers to futuristic space forces and operations (space information warfare, space blockade warfare, space orbit attack warfare, space defense warfare, and space-to-land attacks)—all of which the PLAAF presumably intends to employ in the future.[10]

For now, therefore, integration of air and space clearly remains an aspiration, not a reality. Current publications of the Chinese military focus on space-based information systems to support informationized warfare, although they also refer to "space control" [制天权], which envisions dominance of a space battlefield, including the use of space weapons and attacking ground targets from space.[11] Place hold-

---

[8] PLAAF, 2005, p. 57.

[9] PLAAF, 2005, p. 73. See also Zhang Zhiwei [张志伟] and Feng Chuanjiang [冯传奖], 〈试析未来空天一体作战〉 ["Thoughts on Future Integrated Air-Space Operations"], 《军事科学》 [*Military Science*], Vol. 2, 2006, pp. 52–59, on U.S. "space warfare exercises."

[10] PLAAF, 2005, p. 48.

[11] See Zhang Zhiwei and Feng Chuanjiang, 2006, pp. 15–16, 19.

ers, moreover, seem to have been inserted throughout the *China Air Force Encyclopedia* for further developing and implementing this concept.[12] "Air and space integration," "control of space," or other similar concepts receive mention in the "prospects for the future" section of many major headings, ranging from "Air Force Strategy" to "Air Force Operation Patterns."

However, whether the PLAAF will gain ownership of PLA space assets and missions is uncertain. Currently, most of the PLA's space resources are understood to be held and operated by the General Armaments Department, but this is probably a temporary arrangement, as the General Armaments Department is not an operational service. As Chinese space capabilities have matured and become more operationally relevant, a lively debate over who should ultimately control these assets has emerged. The PLAAF has argued that the unitary quality of the air and space environment, together with the heavy utilization of information derived from space in the USAF and other advanced air forces, suggests that the PLAAF should own China's military space assets and missions and become an air-space force.[13] Unlike the USAF of the past several decades, however, the PLAAF faces stiff competition for the space mission from the Second Artillery. With more experience in ballistics and other areas of space-related physics, Second Artillery officers have claims to greater subject-matter expertise and have advanced arguments for an independent space service (or "space force") to be created after a transitional period in which different elements of it would mature within other relevant services and general departments.[14]

It is difficult to predict which of these organizations, if either, will prevail or whether some compromise solution might be found. We note, however, that, although the Second Artillery is, in many ways, the bureaucratically weaker of the two, the PLAAF's argument for the

---

[12] PLAAF, 2005, pp. 48, 55, 56, 67, 73.

[13] For example, see Cai Fengzhen [蔡风震], Tian Anping [田安平], Chen Jiesheng [陈杰生], Cheng Jian [程建], Zheng Dongliang [郑东良], Liang Xiaoan [梁小安], Deng Pan [邓攀], and Guan Hua [管桦], eds., 《空天一体作战学》 [*The Study of Integrated Air and Space Operations*], Beijing: 解放军出版社 [Liberation Army Press], 2006.

[14] For example, see Yang Xuejun and Zhang Wangxin, 2006.

subordination of space to the air force may be less palatable to other parts of the PLA than the Second Artillery's preference for an independent service. Also notable is the fact that some of the PLA's most prominent military scholars appear to side with the Second Artillery in this preference.[15]

Preparing Simultaneously for the Offensive and the Defensive

According to this "guiding thought for PLAAF construction," the PLAAF should plan to build both offensive and defensive airpower, ensuring capabilities for both in its force structure; organization; training; command, control, communications, computers, and intelligence (C4I) systems; weaponry and platforms; and support and logistics systems.[16] Emphasis on offensive warfare equal to that on defensive warfare is recent; the emphasis was mainly on defensive operations until around 1999. Offensive operations now play a much more important role in PLAAF employment concepts, though defensive concepts continue to be strongly embedded in all descriptions of PLAAF campaigns. China's shift in emphasis from air-to-air to multirole platforms (such as more-recent versions of the J-8, originally an air superiority–only fighter; the acquisition of Russian-made multirole Su-30s; the indigenous development of the J-11B, a multirole version of the Russian-designed Su-27SKs that China coproduced from 1998 to 2004; and indications that the indigenously designed J-10 will be a multirole fighter) and employment concepts exemplifies this increased emphasis

---

[15] For example, see Li Daguang [李大光],《太空战》[*Space War*], Beijing: 军事科学出版社 [Military Science Press], 2001. Uncertainty about which organization will control China's space forces in the future is perhaps not surprising, since deciding where in the PLA space would reside and how it would be implemented is an enormously difficult political task to undertake and deciding to do so could have diplomatic ramifications for a country that publicly claims to oppose the weaponization of space. (For example, the Chinese government's 2008 National Defense White Paper states, "The Chinese government has all along advocated the peaceful use of outer space, and opposed the introduction of weapons and an arms race in outer space." See Information Office of the State Council of the People's Republic of China, *China's National Defense in 2008*, Beijing, January 21, 2009.)

[16] PLAAF, 2005, p. 39.

on offensive operations, as do its calls to attack enemy air forces before they leave the ground.[17]

There do not appear to be "strategic bombing theorists" among PLAAF writers who advocate strategic bombing as an independent and direct route to victory. Nevertheless, PLAAF assessments of the historical development of airpower emphasize the ever-growing importance of airpower, and especially offensive operations, as a decisive form of military power. While defensive operations still receive greater emphasis than they do in the air forces of most Western nations, Chinese theory now emphasizes, "Offensive action is the most basic and most effective form of action in gaining and maintaining the initiative in air campaigns."[18] PLAAF analysts frequently emphasize the independent role of offensive airpower in modern warfare. One set of PLAAF writers assert, "independent, high-tech air raids and counter–air raids will become the basic form of future high-tech, limited warfare." They herald recent military operations, such as the 1999 NATO campaign in Kosovo and, to a lesser extent, the 1991 Gulf War, as cases in which airpower has already played the deciding role.[19]

New Uses of Airpower

The *China Air Force Encyclopedia* defines *airpower* [空中力量] as an overall term for aviation units of air forces, navies, ground forces, "air

---

[17] "SAC J-8," *Jane's All the World's Aircraft*, March 9, 2009; "Sukhoi Su-30 (Su-27PU)," *Jane's All the World's Aircraft*, February 14, 2008; "SAC (Sukhoi Su-27) J-11B," *Jane's All the World's Aircraft*, November 12, 2009; "Sukhoi Su-27," *Jane's All the World's Aircraft*, February 14, 2008; "CAC J-10," 2010. This is the authors' assessment based on a close reading of several entries in PLAAF, 2005; Bi, 2002; and PLAAF, 1994. While entries on air force strategy in PLAAF, 2005, pp. 55–57, discuss both offensive and defensive operations, the PLAAF's discussion of "new uses of airpower," 2005, p. 82, discusses offensive operations first. Within discussion of offensive operations (such as air raids and air offensive operations and campaigns), moreover, the importance of striking first is emphasized. See, for example, PLAAF, 2005, p. 115, under "air offensive combat" ("the objective is an attack on enemy air (water) targets from the air to weaken the enemy") and PLAAF, 2005, pp. 70–71.

[18] Zhang Yuliang, 2006, p. 561.

[19] For example, see Cui et al., 2002.

defense forces" (such as Russia's ProtivoVozdushnaya Oborona, or PVO), and aviation units of SOF.[20] In joint operations, airpower is said to be used for high-speed, in-depth strikes, and to be used first and throughout campaigns to seize control of the skies in support of broader campaign objectives. Airpower is used against key targets, in coordination with other forces. It also is used defensively to protect an air force's ability to conduct air operations, especially air bases, air defense positions, and radar sites, as well as to protect ground and naval operations.[21]

In the encyclopedia's entry on airpower, emphasis is placed on how airpower has changed with the advent of new technology. According to the PLAAF, the use of airpower has changed in a number of ways with a new generation of "informationized air force weaponry," which has advanced air force operational capabilities and created "new" concepts in airpower. These "new" concepts represent aspirations for the PLAAF and areas for future improvement. New concepts for the uses of airpower include the following:

- executing strategic campaign coercion [进行战略战役空中威慑]
- independent and concentrated use of airpower [独立并集中使用]
- conducting joint operations with other services [与其他军种联合作战]
- strategic force delivery [战略兵力投送]
- seizing information superiority and electromagnetic superiority [争夺制信息权和制电磁权].[22]

Military texts repeatedly emphasize these five points, or variations of them, in discussions of air force operations. Of these points, strategic force delivery is probably emphasized the least, except in airborne operations.

---

[20] PLAAF, 2005, p. 81.

[21] Bi, 2002, pp. 140–141, 145–146.

[22] PLAAF, 2005, p. 82.

Other services also participate in air campaigns. In air campaigns, the PLAAF generally is supported by the other services—in particular, receiving naval firepower support to operations close to the coast and conventional missile and Second Artillery support for targets further in the enemy's rear.

Air, Space, Information, and Electromagnetic Superiority

PLA publications assert that the struggle for dominance of the battlefield will increasingly consist of an integrated struggle for air, space, information, and electromagnetic (and even network) superiority.[23] This belief is, of course, closely linked to PLA concepts of informationized warfare and its emphasis on air and space integration, which posits that the struggle for command of air and space will be closely connected. Space is a natural extension of the broadening parameters of air warfare, a location from which weapons could someday be launched, and an important source of information in an era of "informationized" war. Even as integration takes place, PLA writings on the struggle for air superiority illustrate that each component (e.g., air, space, information) of integration will retain a distinct role in the future: While the struggle for command of the air is principally centered on destroying enemy aviation forces, obtaining information control and having a certain level of control of space are "necessary conditions" for obtaining air superiority.[24]

Acquiring Air Superiority

Acquiring air superiority is considered a prerequisite in a variety of operations involving all services. The Chinese term 制空权 means, literally, "control of the air," but is considered essentially equivalent to

---

[23] Of these concepts, network superiority is the least developed. Probably for this reason, there is not a separate entry in the 2005 encyclopedia on network superiority. It therefore is not treated separately in this section, either.

[24] PLAAF, 2005, p. 41.

the U.S./British term *air superiority*.[25] PLA publications state that, by obtaining air superiority, one can restrict enemy air, air defense, and ground forces' operational movements, while ensuring that one's own ground and navy forces have effective cover from the air to carry out their operations. Obtaining air superiority is considered a necessary step in most joint campaigns, ranging from island-landing campaigns and mountain offensive campaigns to the four types of air force campaigns (described in Chapters Five through Eight).[26] As one of several requirements for achieving campaign objectives, however, air superiority is not regarded as the objective of an air campaign, but rather as a means for achieving those objectives.[27]

PLA publications differentiate between strategic, operational, and tactical air superiority. Strategic or "comprehensive" air superiority is command of the air in all stages of combat or across all battlefields or theaters. Operational or tactical air superiority is also called "partial" or "local" air superiority. Operational air superiority is gaining command of the air over the campaign area or at least the most-critical portions of the campaign area, for the entire or most-critical part of the campaign. Tactical air superiority covers a comparatively short period

---

[25] PLAAF, 2005, pp. 39, 41. *Air superiority* is actually considered to be a U.S. and British term but "basically has the same meaning" as the Chinese term (which the PLAAF usually translates into English "as command of the air"). PLAAF, 2005, p. 41.

[26] See Zhang Yuliang, 2006, pp. 311, 313, for island-landing campaigns, and pp. 410–411 for mountain offensive campaigns. The USAF's top-level doctrine publication provides a very similar description of the role and importance of air (and space) superiority:

> Gaining air and space superiority . . . enhances and may secure freedom of action for friendly forces in all geographical environments—land and sea as well as air and space. Air and space superiority provides freedom to attack as well as freedom from attack. Success in air, land, sea, and space operations depends upon air and space superiority. (See U.S. Air Force, *Air Force Basic Doctrine*, Washington, D.C., Air Force Doctrine Document 1, November 17, 2003, pp. 76–77.)

[27] Liu Yazhou [刘亚洲], Qiao Liang [乔良], and Wang Xiangsui [王湘穗], 〈战争空中化与中国空军〉["Combat in the Air and China's Air Force"], in Shen Weiguang [沈伟光], ed., Xie Xizhang [解玺璋] and Ma Yaxi [马亚西], assoc. eds., 《中国军事变革》 [*China's Military Transformation*], 新华出版社 [Xinhua Press], 2003, p. 88. This belief is also present in other, more official, writings.

of time and only part of the battlefield.[28] The PLA does not aim for strategic air superiority. The PLA asserts that air superiority in general and operational air superiority, in particular, are relative.[29] Instead, it aims to achieve enough air superiority to accomplish its campaign or tactical objectives: "one side cannot possibly have comprehensive, absolute air superiority; rather, one has just relative, local air superiority in a definite time period, area (or direction), and to a definite extent."[30] For the PLA, which is likely to face a militarily superior foe in the form of the United States, this would appear to be a realistic attitude toward achieving air superiority.

The PLAAF aviation branch is the primary force in efforts to obtain air superiority, with help from tactical missile forces, army aviation forces, air defense forces, and SOF. Missile attacks, air-to-air combat, and air-to-ground operations are all used to weaken the enemy's air combat and air defense capabilities.[31] The effects of these methods sometimes overlap, so they can be used together or in lieu of one another depending on the targets being struck, the operational space, the stage of the war, and the capabilities of the PLAAF and the enemy.[32]

The PLA prefers to achieve air superiority by attacking the enemy on the ground or water: enemy forces, equipment, bases, and launch

---

[28] For comparison, the USAF describes air superiority as "that degree of dominance that permits friendly land, sea, air, and space forces to operate at a given time and place without prohibitive interference by the opposing force." Air supremacy is described as "that degree of superiority wherein opposing air and space forces are incapable of effective interference anywhere in a given theater of operations" (U.S. Air Force, 2003, p. 77). Thus, the Chinese concept of strategic air superiority corresponds approximately to the U.S. concept of air supremacy, and the Chinese concepts of operational and tactical air superiority correspond approximately to the U.S. concept of air superiority.

[29] Zhan Xuexi [展学习], ed., 《战役学研究》 [*Campaign Studies Research*], Beijing: 国防大学出版社 [National Defense University Press], 1997, p. 307.

[30] PLAAF, 2005, pp. 39–40. Similarly, the USAF states that, while air supremacy "is most desirable, it may exact too high a price. Superiority, even local or mission-specific superiority, may provide sufficient freedom of action to accomplish assigned objectives" (U.S. Air Force, 2003, p. 77).

[31] PLAAF, 2005, p. 40.

[32] PLAAF, 2005, p. 41.

pads used for air raids. Especially at the beginning of a war, the PLA will attack enemy air bases [航空兵基地], ballistic missile bases [弹道导弹基地], aircraft carriers [航空母舰], and warships equipped with land-attack cruise missiles [装载巡航导弹的舰艇] before enemy aircraft can take off or other forms of enemy air strike can be carried out. The intention is to rapidly alter the balance of power in the air. Forces used to launch these attacks will usually include aviation forces, missile forces, and naval forces (submarines and other naval forces would presumably carry out some of the strikes against aircraft carriers).[33]

Fighting the enemy in the air is another means of achieving air superiority, using both air-to-air and surface-to-air operations. Air-to-air and surface-to-air operations can serve as an important means of weakening the enemy, particularly in defensive operations, or when enemy air raid forces and equipment have relatively tough ground- or sea-based defenses and are difficult to attack on the ground or water. The effectiveness of air-to-air operations is said to depend on the quantity and quality of the two sides' air forces and air defenses, as well as their ability to organize and command. Achieving air superiority through air-to-air operations is said to require repeated engagements, and the price to produce visible results will be high.[34]

Another means to achieving air superiority is to carry out air and land attacks to destroy and suppress ground-based air defense systems and air defense command systems.[35] The enemy's command-and-control system is regarded as central to its ability to maintain air superiority, and the enemy's air defenses are regarded as the greatest threat to aviation forces before they gain air superiority. Striking both is therefore considered an effective way to gain command of the air. When complete destruction of enemy command-and-control systems and enemy ground-based air defense systems proves to be difficult, PLA publications advocate destroying or suppressing targets in the vicinity of friendly lines of flight to establish air corridors that can be defended

---

[33] PLAAF, 2005, pp. 39–40.

[34] PLAAF, 2005, p. 40.

[35] PLAAF, 2005, p. 40; Zhan, 1997, pp. 310–312.

against attack.[36] The PLA writings on achieving air superiority also advocate destroying enemy air industries and aviators' training bases to cut off the enemy's ability to supply aircraft and pilots.[37]

Finally, defensive operations are an important component of air superiority throughout a campaign. The continued emphasis on defensive operations during operations to gain air superiority may reflect the PLA's view that air superiority is relative. Defensive operations emphasize passive measures to avoid strikes against aviation forces on the ground: Between battles, aviation forces are instructed to disperse deployments and camouflage themselves. They also are instructed to defend operational aircraft to prevent them from being struck on the ground by the enemy.[38]

Acquiring Space Superiority

According to PLA publications, space superiority is when one party engaged in combat has command of outer space for a definite period of time and within defined parameters. Objectives are said to include gaining superiority in space, protecting one's own ability to maneuver in space, and depriving the enemy of its ability to maneuver in space. In future warfare, space superiority is expected to be crucial for controlling the ground, naval, and air battlefields. To gain space superiority, offensive and defensive weapon systems will be deployed on the ground, air, sea, and space. The "space force" [天军], combined with other services, will lead operations, which can include space information warfare, "space blockade warfare," space orbit attack warfare, space defense warfare, and space-to-land attacks.[39]

The PLA's vision of space warfare is still vague and theoretical, though a movement toward integration of space is portrayed as a characteristic of future warfare.[40] The *China Air Force Encyclopedia*'s entry

---

[36] PLAAF, 2005, p. 40.

[37] PLAAF, 2005, p. 40.

[38] PLAAF, 2005, pp. 40–41.

[39] PLAAF, 2005, pp. 48–49.

[40] PLAAF, 2005, pp. 47–48.

on space warfare, for example, discusses only U.S. and Russian space warfare concepts and capabilities, rather than describing universal principles that might apply to future Chinese capabilities or operations.[41]

Acquiring Information Superiority

In struggles for information superiority, the goal is to control information on the battlefield, allowing the battlefield to be transparent to one's own side but opaque to the enemy. The ultimate goal is to control the enemy's information systems. As with air superiority, information superiority can be strategic, operational, or tactical; it can be complete or partial; and it generally occurs in a specific time and space.[42] Methods for achieving information superiority include achieving electromagnetic superiority through interference, achieving network superiority through network attacks, using firepower to destroy the enemy's information system, and achieving psychological control. Concepts for achieving electromagnetic superiority are described in much greater detail than concepts for achieving network superiority, firepower attacks on enemy information systems, or achieving psychological control, and are discussed in the next section. Network attacks include using computer viruses and computer invasions to weaken and break enemy computer network systems' functionality while protecting friendly computer network systems' operations. Firepower methods can include laser, radio frequency [射频], particle beam [粒子束], and other direct attacks on enemy information systems. Psychological control can be achieved using psychological propaganda, deceit, and threats sent through television and radio broadcasts, the distribution of leaflets, and electronic mail.[43] The ultimate goal is to influence the psychology of individuals, governments, organizations, and other bodies, impairing their objectivity and affecting their actions.[44] Finally, strict information defense—countering electronic surveillance,

---

[41] PLAAF, 2005, p. 49.

[42] PLAAF, 2005, p. 49.

[43] PLAAF, 2005, p. 50.

[44] PLAAF, 2005, p. 50.

blocking enemy computer viruses and hackers, and resisting electronic interference—is key to ensuring that PLAAF information continues to have secure and stable operations.

Acquiring Electromagnetic Superiority

In high-tech wars, having electromagnetic superiority is viewed as necessary for securing space, air, and sea superiority, and, in recent sources, electromagnetic superiority in the air and "outer space" is considered "key" to electromagnetic superiority overall.[45] While acquiring electromagnetic superiority is described as a subset of acquiring information superiority, it is treated as a distinct operation in PLA publications, in contrast to achieving network superiority or the use of firepower to achieve information superiority. Electromagnetic superiority is defined as "electromagnetic control by one side engaged in battle for a specified period of time and within specified parameters." As with the other "superiorities," the PLA does not expect to acquire absolute, or strategic, electromagnetic superiority. It seeks to deny the enemy capabilities provided by electronic equipment while it retains dominance during the period necessary to successfully conduct operations. Electromagnetic superiority is divided into strategic (across the entire theater for a sustained period), campaign (during the entire or important parts of a campaign), and tactical (during specific operations) superiority.[46]

During air force campaigns, EW is considered an important component of information warfare operations and principally an offensive action.[47] EW is said to usually occur at the beginning of operations and to possibly last throughout the conflict. It is said to take place over a broad area, to use many different methods (including both soft and hard measures), to be very technical, and to face a complex battlefield in which movements are rapid and command coordination can be

---

[45] PLAAF, 2005, p. 50.

[46] PLAAF, 2005, p. 50; see also Zhan, 1997, pp. 297–305.

[47] Zhan, 1997, p. 300.

unpredictable.[48] Recent historical examples given of the effective use of EW include Israel's 1982 electronic attack against Syrian forces.

> Under the command and control of their early warning aircraft, [Israel] first carried out a large-scale electronic attack to achieve electromagnetic superiority, and then, with the assistance of electronic war, within six minutes had destroyed nineteen of the Syrian military's SAM-6 missile bases.[49]

Other examples cited are the use of EW in the Gulf War (1991) and Kosovo (1999), in the latter case including conventional electromagnetic pulse bombs that are said to have caused widespread paralysis to Yugoslavia's electronic information networks (communications, computers, and other electronics).[50]

Methods for obtaining electromagnetic superiority are said to include electronic attack and electronic defense. In electronic attack, soft kill measures include electronic interference and electronic deception. Hard kill measures are said to include "antiradiation destruction" [反辐射摧毁], "electronic weapon attack" [电子武器攻击], "firepower destruction" [火力摧毁], and attacks against the enemy's electronic installations and systems to destroy, weaken, or obstruct the enemy's use of the electromagnetic spectrum. Weapons used to conduct electromagnetic attacks are said to include electromagnetic pulse bombs and high-power microwave weapons. Electronic defense is simply defending against enemy electronic and firepower attacks.[51]

In EW, establishing a plan is said to come first. With a plan in place, the campaign commander can determine the deployment and composition of EW forces to be used—including those residing within the services and at the campaign and tactical levels. Coordination of communications, radar, and technical reconnaissance is said to need to be organized effectively around the unit with the main mission.

---

[48] PLAAF, 2005, p. 105.

[49] PLAAF, 2005, p. 50.

[50] PLAAF, 2005, p. 50.

[51] PLAAF, 2005, pp. 50–51.

According to PLA publications, each unit's responsibilities, operational methods, airspace, territory, methods of coordination, and needs for coordination must be well known, and each unit must be able to adjust rapidly and regain contact if it is cut off.[52]

The primary targets of EW are said to include command, control, communications, and intelligence systems.[53] In an era of information-ized warfare, EW has developed and expanded to include almost every aspect of operations at all levels: in space, in the air, on the ground, and at sea.[54] Missions of EW are said to include electronic surveillance, interference, suppression, destruction, deception, and defense. Anti-radiation missiles are described as a preferred weapon for suppression. Maintaining control of one's own electromagnetic spectrum is said to be the goal of electronic defense; achieving this can rely on procedural actions, such as declaring a blackout on wireless communications or using coded dispatches. It can also include more-active measures to counter electronic interference.[55]

As mentioned earlier, the PLAAF views future electromagnetic struggles as closely tied to computer network warfare; in some cases, struggles for electromagnetic superiority and network superiority will emerge as the main focus of the battle for information superiority. In addition, the use of the electromagnetic spectrum will continue to expand and technology to wage EW will become more advanced and diverse.[56]

---

[52] PLAAF, 2005, p. 105.

[53] Zhan, 1997, p. 300.

[54] Zhan, 1997, p. 298.

[55] Zhan, 1997, pp. 301–302.

[56] PLAAF, 2005, p. 51.

Three Modes of Air Combat Used to Achieve Operational or Campaign Objectives

The *China Air Force Encyclopedia* outlines three major types of air combat operation: air-to-air combat, air-to-surface combat, and surface-to-air combat.[57] These types of combat operation can occur within any of the four types of air force campaigns described in Chapters Five through Eight.

Air-to-Air Operations

Air-to-air operations are an area of traditional emphasis for the PLAAF, in part because, in the past, the PLAAF has been very defensively oriented. The PLAAF seems to be moving away from emphasizing air-to-air operations, emphasizing instead operations to gain air superiority by attacking enemy airfields and controlling the enemy on the ground before resorting to fighting the enemy in the air.[58] This may be in part because the PLA now believes that air-to-air operations exact a high price to conduct effectively.[59]

Air-to-air operations can be used in any of the four types of PLAAF campaigns, including operations within any of those campaigns to defend PLAAF forces or to gain air superiority. Air-to-air operations are used to counter enemy aircraft flying either defensively to intercept incoming PLAAF aircraft or offensively to attack Chinese territory. Air-to-air operations are considered an important method for destroying enemy aircraft, gaining air superiority, and covering one's own aircraft, troops, and troop operations.[60] They are used to resist

---

[57] Under the entry on "Air Force Tactics" [空军战术] in PLAAF, 2005, p. 108, the three types of tactics listed and described are air-to-air combat [空中战斗], air-to-surface combat [空地战斗], and ground-to-air combat [地空战斗].

[58] Zhan, 1997, pp. 310–312. Some forward-leaning theorists believe that air-to-air operations should be deemphasized even more than they have been. For example, Liu, Qiao, and Wang, 2003, pp. 87–88, argue that the PLAAF needs to break through "air combat" [空战]–centered thought and work toward the idea that the main type of combat is "air raids" [空袭].

[59] Zhan, 1997, p. 310; PLAAF, 2005, p. 40.

[60] PLAAF, 2005, p. 110.

enemy air attacks against friendly ground targets; obstruct the enemy from approaching or entering the airspace above important friendly political, economic, and military targets; and, in the process of the campaign, to strive for battlefield air superiority. In an offensive air campaign, they are used when the enemy air force bases have strong defenses, enemy aircraft have secure shelters, and it is difficult to destroy aviation forces by attacking their airfields.[61] The effectiveness of air-to-air combat depends on how many aviation and air defense forces each side has, as well as their ability to organize and command.[62]

It is worth noting that, in air-to-air operations, as in other operations, the PLAAF emphasizes surprise and other methods that give an inferior force an advantage. Air-to-air combat generally is not the preferred method of operations: It is used mainly when enemy aircraft cannot be destroyed on the ground and the PLAAF needs to press on with an air-to-air engagement in order to meet its campaign objectives. Air-to-air combat was probably emphasized more in the early 1990s, when the PLAAF's *Study of Air Force Tactics* was published.[63] At that time, surprise and positioning were emphasized in writings on air combat even more than in more-recent publications.

While a flexible mix of different tactics is suggested in different combat situations, current writings on air-to-air combat suggest that striking from a distance with long-range weapons is preferable to fighting at close range. Even with these changes, what is interesting is that the four phases of combat in air-to-air operations outlined in military texts (searching, engaging the enemy, attacking, and retreating from combat) have remained unchanged from 1994 to 2005.

The Stages of Air-to-Air Combat. There are four stages of combat in air-to-air operations: (1) search, (2) engagement, (3) attack, and (4) withdrawal from combat.[64]

---

[61] PLAAF, 2005, p. 99.

[62] Zhan, 1997, p. 310; PLAAF, 2005, p. 40.

[63] PLAAF, 1994.

[64] This section is drawn from PLAAF, 2005, p. 111, for the basic principles; most of the detail is derived from PLAAF, 1994, pp. 120–127. Both PLAAF, 2005, p. 111, and PLAAF,

First Stage: Search [搜索]. This stage entails using visual means or technical equipment to detect and discern the status of any movement in the air from the time the aircraft takes off until the time it finds the target. As early as 1994, PLA internal writings noted that a basic principle to which to adhere during air combat is that, if a pilot sees the enemy first, the pilot fires first to protect him- or herself, regardless of the mission that he or she is carrying out.[65] During the search phase, the command center uses radar and electronic interference to probe. The pilot also uses his or her own visual and technical means, which may be more timely, accurate, and rapid than those of the command center, to search the area in case the command center has not picked everything up, especially if there is extensive electronic interference. The command center in theory allows the pilot some independence to probe and do early warning. For example, pilots are advised to use multidirectional probes, though mostly focused on the direction from which the enemy is likely to fly.[66]

Second Stage: Engagement [接敌]. According to PLA writings, it is important to find a good position for attack from the start of flight operations until the pilot is in position to attack the target. Positioning remains an important concept for the PLAAF today; it was even more important in 1994, when the PLAAF did not have advanced fighters and emphasis was on positioning to make up for having inferior aircraft: "the aircraft with the advantageous position has a better survivability rate even than an aircraft with better capability."[67] Nevertheless, with continued emphasis on position, it is likely that some tactics from 1994 still apply today, such as using camouflage, sneaking up on the enemy from the enemy's blind spots or weak surveillance areas (even the tactics book, published in 1994, notes that this is difficult to do with modern technology), and then immediately shooting at the

---

1994, pp. 120–127, describe the same four phases. Because PLAAF, 1994, is a book on tactics, however, it contains more detail, described here.

[65] PLAAF, 1994, p. 120.

[66] PLAAF, 1994, pp. 120–123.

[67] PLAAF, 1994, p. 123.

enemy. Since the widespread use of all-aspect missiles began, shooting from the rear of the aircraft is not an adequate tactic.[68]

Third Stage: Attack [攻击]. This stage involves using firepower to destroy enemy aircraft. This is the decisive stage of combat. It begins with the pursuit and lasts through the lock-on to the target, shooting, and withdrawal from attack.[69] Striking first and accurately is considered important: It is difficult to strike again if the pilot misses on the first strike because the enemy will strike back. Because speed and accuracy are both important—but one makes the other difficult—the PLAAF advises trying to surprise the enemy so there is more time to accurately aim and shoot. Shooting while retreating also can be effective.[70]

A book on the use of missiles in combat[71] has a useful section on air-to-air combat that sheds light on PLAAF tactics. The PLAAF would choose different positions or formations based on its level of confidence in its capabilities and in different types of combat. The following discussion is based on Ge, 2005, pp. 99–108.

Single-Aircraft Missile Attack Tactics. Four types of tactics are described for single-aircraft attacks. The four types of tactics vary according to the capabilities of the aircraft in relation to the enemy's aircraft, the angle from which the aircraft will attack, and the distance from which it attacks (especially if it has over-the-horizon capabilities):

- *Meet head on; seek to attack first* [迎头进入，抢先攻击]. This is the main tactic suggested for attacking with aircraft that have weapon systems that are superior to those of the enemy. This tactic works as follows: Once the command center spots the enemy aircraft at a distance flying head-on, the pilot, acting on the order of the command center, turns on the aircraft radar, searches, detects, acquires, and identifies the target. If, however, the radar is turned on too early, the pilot risks exposing himself or herself (therefore,

---

[68] PLAAF, 1994, p. 124.

[69] PLAAF, 2005, p. 111.

[70] PLAAF, 1994, pp. 126–127.

[71] Ge Xinqing [葛信卿], 《导弹作战研究》 [*Research on Missile Operations*], Beijing: 解放军出版社 [Liberation Army Press], 2005.

a good defensive capability is an advantage after turning on the radar); if the radar is turned on too late, it is easy to miss a good opportunity. The comparative capability of the enemy's radar is a critical factor in the success or failure of this tactic, as is a high kill capability: If one misses the first time, it is difficult to find a second chance to strike.

- *Approach from below; seek to attack first* [仰头进入, 抢先攻击]. This tactic is suggested for use when weapon system capabilities are inferior to those of the enemy aircraft but there is some over-the-horizon air combat capability.

- *Avoid the front aspect; attack from the side (or rear)* [避开正面, 侧（后）攻击]. This tactic is a description of often-cited PLAAF strategies to attempt to surprise the enemy by arriving within shooting range by approaching through its blind spot. Upon receiving a command from the command center, the aircraft closes in on the enemy aircraft, maintaining a certain distance (specifically avoiding a head-on angle). When within range of the enemy radar, the pilot turns on the aircraft's radar and then attacks from the side. By approaching at a 120-degree angle, the aircraft takes advantage of the enemy radar's shortest and weakest point, making it more difficult for the enemy to probe and lock onto a target. The pilot coordinates actions closely with the command center and requests precise guidance if the situation changes significantly during this kind of attack.

- *Combine aircraft maneuverability and jamming; evade when distant, and attack when close* [机动与干扰相结合, 远避近攻]. This tactic is used mainly when one does not have an over-the-horizon capability. There are two ways to carry this out: with a "tail" maneuver, coming up quickly from behind the enemy, or attacking from below at a 180-degree angle. Another is finding the "optimal azimuth" and moving to about a 70- to 80-degree angle when close to the target.

Group Missile Attack Tactics. In addition to combat between solitary aircraft, suggestions are also provided for using missiles in air-to-air battles between whole formations of aircraft. Again, four tactics are suggested, this time oriented on a group of aircraft:

- *Cooperation between real and feint forces; coordinated attack* [主佯 配合, 协调攻击]. This tactic aims to deceive the enemy, leading it to confuse the decoy forces for the real forces. It aims to use the feint to hide the real direction of the attack.
- *Approach from two directions; continuously or simultaneously attack* [双方进入, 同时或连续攻击]. This tactic divides and diverts the enemy, since the enemy cannot go after both groups at once, particularly because aircraft fly toward the enemy at an angle (70–80 degrees) that is difficult for the enemy radar to detect. The PLAAF can closely coordinate and fly in one formation when far from the enemy, but, as PLAAF aircraft approach, they split off into two formations at 70- to 80-degree angles from each other, and then turn again at a 70- to 80-degree angle to attack the enemy from either side. The formations should fly at relatively low altitude. Coordination is complex and requires adhering closely to instructions from the command center.
- *Full-force attack* [权力攻击]. This tactic has the advantage of being a continuous attack; it also uses two columns, one after the other, making it difficult for the enemy to detect the second column. The first column attacks first. The second column attacks immediately after the first column, maintaining continual, strong firepower. This tactic is premised on making it difficult for the enemy to defend itself due to mistakes in its judgment, presuming that, given a relatively long lag time between the arrival of the two columns into enemy radar coverage, the enemy has not detected the second column when it has already detected the first column.
- *Prepare an ambush in advance; attack suddenly* [预先设伏, 突 然攻击]. In this tactic, participating formations break into two parts: one part to lure the enemy and the other to ambush the enemy. This kind of attack occurs when the ambush force is at an advantage and over-the-horizon warfare offers an advantage. Camouflaging the ambush force and decoy force are both important. Leaving the air base at a low altitude, radio silence, elec-

tronic jamming, the use of radar to probe for blind spots, and camouflaging intentions are some methods for achieving this.

Fourth Stage: Withdrawal from Combat [退出战斗]. This is the concluding phase of the air-to-air operation. It begins with the completion of the attack and lasts until the aircraft has landed. In over-the-horizon fights, attacking from a distance with weapons and missiles that can automatically lock on and attack targets simplifies engagement with the enemy, attack, and withdrawal.[72] The withdrawal phase is considered just as dangerous as any of the other phases of combat, and it is recommended that aviators use camouflage or retreat under cover. If a group of aircraft is retreating, then those with no ammunition or little fuel go first; the command aircraft directs aircraft below it; and covering aircraft protect aircraft below. Aircraft providing cover retreat last.[73]

Modern technology is said to have changed some aspects of air-to-air combat. Air-to-air combat is said to now involve more-mature beyond–visual range technology, more-destructive firepower, more-lethal aircraft, more-integrated use of electronic countermeasures in air operations, diminished effects of weather and time of day, and both a greater threat and greater assistance from ground-based air defense systems.[74]

Air-to-Surface Operations

Air-to-surface operations seem to be the preferred modus operandi for PLAAF offensive operations, so much so that the first listed objective of "air offensive combat" [空中进攻战斗] is to "attack enemy ground (or water) targets" (the other two are to weaken the enemy's war potential and to carry out battlefield interdiction and air firepower support).[75] A preference for air-to-surface operations is simultaneously an aspiration

---

[72] PLAAF, 2005, p. 111.

[73] PLAAF, 1994, p. 127.

[74] PLAAF, 2005, p. 111.

[75] PLAAF, 2005, p. 115. The entry for "air-ground combat" is limited to a one-sentence definition and refers to "air offensive combat" for more detail.

and a growing reality for the PLAAF. Though air-to-air operations also play an important role in air offensive combat, air-to-surface operations are considered more effective, less costly, and less reactive than air-to-air operations.[76]

In the past, the PLAAF emphasized air-to-air operations because of the defensive nature of PLAAF operating concepts, and the PLAAF remains primarily a defensive air force in comparison to the USAF. Nonetheless, its acquisition of Su-30s and other platforms with good air-to-surface capabilities and its increased interest in the eventual use of airpower to achieve strategic objectives all reflect a PLAAF that is trying to become more offensive and be capable of taking the initiative.

The principal air-to-surface operations, discussed in more detail below, are air strikes, air raids, deep air strikes, advance firepower preparation, and close air firepower support.

Air Strikes [空中突击 **or** 航空兵突击]. In USAF usage, *air strikes* tends to refer to attacks against one target or target set. As they are for the USAF, air strikes in the PLAAF's lexicon do tend to be limited to one operation, a fairly focused effort at striking a target or group of targets from the air. The PLA's use of this term is not as strict, however. While *air raids* [空袭] (discussed below) tends to refer to larger-scale operations in Chinese, occasionally, the two terms are used interchangeably.

Air strikes are defined as "aviation forces entering into combat action from the air, using bombs, or attacks against land or sea targets."[77] These operations usually use bombers, strike aircraft, and attack helicopter forces, working along with other air forces (such as cover forces). Forces involved can range from single aircraft to formations. Combat methods can include concentrated attack, continuous attack, and simultaneous attack from a low, medium, or high altitude, and can be done using a "level attack" [水平攻击], "gliding attack"

---

[76] For example, in its discussion of "methods" for achieving air superiority, the *China Air Force Encyclopedia* (PLAAF, 2005, p. 40) remarks that attempting to achieve air superiority in the air (using air-to-air and surface-to-air operations) requires "payment of a relatively high price to be effective" [付出较大代价才能显现效果].

[77] PLAAF, 2005, p. 123.

［下滑攻击］, "diving attack" ［俯冲攻击］, or "rising attack" ［上仰攻击］.[78]

Elements of an attack include approaching the target, aiming and launching the missile or bomb, and dropping bombs and shooting missiles. According to the *China Air Force Encyclopedia*, to achieve this effectively, one needs to adopt appropriate tactics; concentrate the use of force in the main direction of the attack and time the attack well; attack vital targets; achieve surprise through the use of electronic interference and attack on a variety of targets; choose the right weapons and corresponding tactics, azimuth, and formation; strike correctly the first time; and closely coordinate between the attack and support forces.[79]

As with most air operations, the PLAAF believes that air-to-surface strikes are evolving with the introduction of new technology. With the development of air technology and the appearance of multi-role aircraft, aircraft previously in an air-to-air combat role will also be able to attack ground targets, and attack aircraft will be more able to defend themselves. Information warfare, network warfare, EW, stealth technology, PGMs, coordination between several different types of aircraft, and attack from various altitudes and depths also will affect future air-to-surface operations.[80]

Air Raids ［空袭］. An air raid is an attack against enemy surface targets from the air. As mentioned earlier, air raid operations tend to be larger in scope than air strikes. Missions include destroying targets in the enemy's rear, weakening the enemy's military might and wartime potential, weakening or suppressing forces and weapons on the battlefield, and supporting friendly forces operations. Air raids can be nuclear or conventional, and, depending on the scope, can be strategic ［战略空袭］, operational ［战役空袭］, or tactical ［战术空袭］. Modern air raids tend to be sudden, destructive, and broad in scope.

Air raid operations, like other operations, require "meticulous planning and full preparation" ［周密计划, 充分准备］, the use

---

[78] PLAAF, 2005, p. 123. Each of these types of attack is also discussed in its own entry in the *China Air Force Encyclopedia*.

[79] PLAAF, 2005, p. 123.

[80] PLAAF, 2005, p. 123.

of surprise, and concentrating the forces to carry out strikes on vital points [集中兵力, 实施重点突击]. In addition, air raid operations include some specific requirements for force composition and division of missions among forces and services. Forces in an air raid should be deployed for an offensive posture. The formation should include a strong strike group [突击集群] as the central force, as well as a suppression group [压制集群] and support group [保障集群] to ensure the strike group's successful completion of the mission.

Coordination between the forces should be well organized. For example, if ballistic missile forces or guided-missile submarines and air forces are both are designated to strike the same target, the ballistic missile forces or guided-missile submarine forces strike first, and the air force strikes afterward. When attacking different targets in the same target system, the ballistic missile forces and guided-missile submarine forces strike targets with stronger enemy air defenses or larger surface areas. The air force, on the other hand, strikes targets with weaker air defenses and smaller surface areas. This arrangement plays to the strengths of the forces involved.[81]

Employment concepts for air raids are in transition. The volume of official and unofficial documents on air raids indicates that extensive thought is being devoted to the concept of air raids.[82] This is due to rapid changes in technology as well as the view, stated in the *China Air Force Encyclopedia*, that "air raids will merge and integrate with space raids." Advances in technology include more-advanced weapons, greater value placed on information control, and aircraft equipped with stealth technology, jamming and antijamming technology, PGMs, ALCMs (improving the ability to conduct long-distance operations), and improvements in the ability to attack targets in all weather and from all positions.

---

[81] PLAAF, 2005, pp. 70–71.

[82] Examples include Cai et al., 2006; Cai Fengzhen [蔡凤震] and Tian Anping [田安平], eds.,《空天战场与中国空军》[*Air and Space Battlefield and China's Air Force*], Beijing: 解放军出版社 [Liberation Army Press], 2004; and Li Rongchang [李荣常] and Cheng Jian [程建],《空天一体信息作战》[*Integrated Air and Space Information Warfare*], Beijing: 军事科学出版社 [Military Science Press], 2003.

Deep Air Strikes [纵深空中突击]. Deep air strikes are strikes against targets or target sets in the enemy's rear in support of army or navy forces. Attacks are mainly carried out against the enemy's in-depth campaign and tactical targets, weapons, transportation systems, and logistics supply system.[83]

In its discussion of mountain offensive campaigns, the most recent version of *Study of Campaigns* offers a good picture of how a deep air strike would work. In this particular description, deep air strikes would be one part of a comprehensive firepower attack that would include strikes across the enemy's entire depth, frontline attacks on key targets, in-depth precision strikes against targets, and isolating the battlefield. The "in-depth precision strike" component represents a deep air strike, though the PLAAF would play a role in each of these operations. The in-depth precision strike operation is usually carried out at the same time as frontline attacks. The air force is the principal actor, assisted by army helicopters and uses long-range precision attack weapons, such as PGMs, to attack and destroy the enemy's in-depth command-and-control systems, EW systems, and other important targets. When executing an in-depth precision attack, responsibilities are divided as follows: Air force attack aircraft focus primarily on enemy command and control and "information confrontation" systems [指挥控制与信息对抗系统]. Bombers carry out precision attacks against enemy logistics facilities, bridges, tunnels, and other transportation hubs. Army aviation armed helicopters and guided artillery mainly make precision attacks against enemy in-depth armored vehicles and radars. In the enemy's rear, with the assistance of SOF, they also actively use laser-guided artillery to carry out precision attacks against enemy small-scale point targets [小型点状目标]. Strike forces of the Second Artillery's conventional missile units carry out precision attacks against important enemy fixed installations, such as electronic installations and underground facilities.[84]

Advance Firepower Preparation [航空火力准备]. The most essential goal of advance firepower preparation is to suppress enemy

---

[83] PLAAF, 2005, pp. 126–127.

[84] Zhang Yuliang, 2006, pp. 414–415.

air defenses, particularly prior to a ground or air offensive operation but also prior to airborne forces' landing and before counterattacks in defensive campaigns. Air firepower preparation is defined as an air firepower strike on enemy targets before army or navy forces attack and includes "preliminary air firepower preparation" and "direct air firepower preparation." Its goal is to attack targets with weak resistance to facilitate ground or naval efforts to penetrate enemy defenses: enemy tactical missiles, groups of tanks, artillery emplacements, support hubs, prepositioned units, and command units. The attack is usually carried out by forces that are responsible for tactical missions, though long-distance bombers also participate as needed. While preliminary air firepower preparation strikes occur several hours or even days before an offensive strike (or other operation), direct firepower support occurs just minutes before launching the attack.[85]

Close Air Firepower Support [近距航空火力支缓]. Close air firepower support usually is implemented in support of other services' offensive or defensive operations by attacking the enemy's near and rear area battlefield targets.[86] Close air firepower support is viewed as the most complicated kind of air-to-surface operation due to the large distance between air deployments and the location of the ground forces (there can be a lag between when ground forces need support and when the air forces arrive on the scene) and proximity of friendly ground forces to the air strike targets (it is easy to accidentally attack friendly forces). For these reasons, close air firepower support requires air forces to respond rapidly, preferably by deploying toward the front or even being put on strip alert, and coordinate any firepower with ground troop operations. The air forces must correctly distinguish friend and foe (including decoys) and receive timely and accurate battlefield information. A "target guidance small group" [目标引导小组] is often dispatched to the ground forces receiving support to help call in and direct air force attacks.[87] Again, coordination with the ground forces is

---

[85] PLAAF, 2005, p. 126.

[86] PLAAF, 2005, pp. 104–105.

[87] A target guidance group is responsible primarily for determining the course, distance, timing, and other parameters that relate aircraft to intended targets; for guiding PLA aircraft

crucial, but procedural precautions are considered necessary: "to avoid mistakes, the aviation forces' targets should be some regulated safety distance from the [friendly] forces."[88] The PLAAF notes that technology will improve direct firepower support, but does not specifically mention the role of technology in increasing coordination in command and control and in distinguishing friend and foe.[89]

Different bombing methods are recommended for different types of operations. Against relatively large three-dimensional targets that are wide or high, such as train stations and warships, level bombing is recommended. Against narrow targets, such as bridges, glide bombing [下滑攻击] is recommended. In strikes against enemy artillery and tactical ballistic missiles, the PLAAF recommends concentrating forces in a strike against the most-threatening enemy forces. Against tanks and armored vehicles that are moving relatively quickly, striking roads that pose the greatest threat to friendly forces is recommended. The *China Air Force Encyclopedia* notes that PGMs allow more-flexible modes of attack against various targets.[90]

Finally, the PLAAF warns against the danger of enemy air defenses and other forces during close air firepower support operations. Tactics to reduce losses and achieve operational goals can include attacking enemy air defenses to paralyze them. Short of this, several measures can reduce the effectiveness of enemy air defenses. One method is to fly at low or extra-low altitude to increase the likelihood that enemy radar will detect attack forces late or encounter difficulties tracking them to provide targeting information for air defenses. Another method entails flying in small formations dispatched in several groups that arrive at the scene from several different directions and using different altitudes

---

toward enemy targets; for assessing the results of air strikes; for assisting aircraft in correcting bombing and targeting errors; and for assisting the army, navy, and air force in distinguishing between friendly and enemy aircraft (PLAAF, 2005, p. 168).

[88] PLAAF, 2005, p. 126.

[89] PLAAF, 2005, p. 128.

[90] PLAAF, 2005, p. 128.

to disperse enemy air defense firepower. Finally, some PLAAF forces can suppress enemy air defenses to protect the main force.[91]

Surface-to-Air Operations

Although surface-to-air operations have historically proven to be some of the PLAAF's most effective operations (at least in terms of number of enemy aircraft destroyed), PLA strategists believe that improved reconnaissance and air strike technologies confront these operations with new challenges. Nevertheless, surface-to-air operations remain important in air offensive, blockade, and, especially, air defense campaigns, and Chinese sources identify a variety of measures designed to ensure their continued relevance. Since 1985, Chinese surface-to-air tactics have moved away from their earlier focus on dealing with large-scale air raids to addressing the challenges posed by medium- and small-scale raids executed by aircraft armed with precision stand-off weapons.[92]

Here, we limit our discussion of surface-to-air operations primarily to SAM operations and do not discuss AAA operations in detail.[93]

Principles of Surface-to-Air Tactics. The principles of surface-to-air tactics include (1) taking the initiative and exercising constant vigilance; (2) the concentrated use of assets and their deployment in mixed packages; (3) emphasis on self-defense, agility, and flexibility; (4) concealment and suddenness; (5) active cooperation and close coordination between elements; (6) combined use of different tactical techniques; and (7) strict security and comprehensive support.[94] As a unified whole, these principles highlight PLAAF commanders' desire to avoid placing their ground-based antiaircraft assets in a reactive posture, but they also reflect an inherent tension between the demands

---

[91] PLAAF, 2005, p. 128.

[92] PLAAF, 2005, p. 133. Other trends include the integration of air defense and space defense capabilities, the integration of hard and soft kill capabilities, and the development of fully automated command-and-control systems.

[93] For a discussion of Chinese AAA operations and tactics, see PLAAF, 2005, pp. 139–145.

[94] PLAAF, 2005, p. 133.

of defending fixed targets and conducting dynamic defense of larger zones.

Combat Disposition [战斗部置] **and Formations** [战斗队形]. SAM units can be arranged in several types of basic dispositions: circular [环形部置], fan-shaped [扇形部置], dispersed linear [宽正面 or 线形部置], or massed [集团部置]. As might be expected, the choice of a specific disposition will depend on a variety of factors, including the operational plan, the mission of the unit in question, the nature of the target being defended, the nature of the terrain, the type of weapon systems deployed, the circumstances of adjacent air defense units, the type of offensive air systems deployed by the enemy, the possible direction of attack, and enemy strength and tactics.[95]

The *China Air Force Encyclopedia* suggests that firing elements will normally be deployed in either a circular or a fan-shaped disposition.[96] When defending against enemy elements attacking from multiple directions, a circular disposition should be adopted. When defending against enemy aircraft attacking from a single or primary direction, a fan-shaped disposition will be more appropriate. The command post will normally be deployed near or to the flank of the firing position's guidance radar. Target acquisition radar positions will also be deployed to one side of the firing position, while support and missile storage areas will be to the rear or rear flank of the firing position (and in a position facilitating the movement and resupply of missiles and where conditions favor concealment).

The main strength—and the most-advanced systems—should be maintained around the most-important defended targets and deployed toward the most likely direction of attack. Elements should be arrayed close to or beyond the "mission completion line" of the attacking enemy weapons. Different types of missile and gun systems should be deployed in mixed packages, with high-, medium-, and low-altitude systems and long-, medium-, and short-ranged systems deployed together for mutual support. Dispositions should be arranged to facilitate command and mobility, and they should combine fixed

---

[95] PLAAF, 2005, p. 133.

[96] PLAAF, 2005, p. 134.

with dynamic deployments. Heavy emphasis is also placed on concealment and camouflage.[97]

Surface-to-Air Missile Positions [地空导弹兵阵地]. SAM positions consist of several component sites: firing positions, technical support positions, target acquisition radar positions, and command post positions. In determining the arrangement of these positions, six criteria are listed: (1) the demands of the operational plan; (2) selection of terrain that enables the full use of firepower; (3) ability to establish effective command and communication links; (4) compliance with electromagnetic compatibility and noninterference standards; (5) ease of constructing positions and camouflage; and (6) effective access to routes in and out of positions and the facilitation of mobility.[98]

Positions are defined as either primary (literally, "basic" [基本]) or reserve [预备]. Primary positions are the main defensive positions for a given unit, while reserve positions are those into which units deploy when on the move. Normally, each firing unit should have two or three reserve positions. SAM positions are also differentiated by the level of engineering preparation: (1) Most primary positions are permanent positions with permanent underground facilities for weapons and personnel. They are normally located around axes of approach to important strategic targets. (2) A second category of positions has some protective capability and is at least partly permanently maintained. These positions have some underground and some partially underground works. (3) Simple positions include some partially underground facilities and protective structures and employ all methods of camouflage. (4) Finally, field positions have temporary shelters for troops and equipment. Given the demands of air defense in modern war, there is a greater need for mobility (see next section) and, consequently, "a need to substantially increase work on reserve positions, as well as the ability to conceal positions."[99]

Maneuver Operations [机动作战]. As the emphasis on surface-to-air operations has moved from fixed-point defense to a more dynamic

---

[97] PLAAF, 2005, p. 133ff.

[98] PLAAF, 2005, p. 134.

[99] PLAAF, 2005, p. 134.

defense of zones or areas, mobility has become increasingly important. The *China Air Force Encyclopedia* defines mobility operations as "operational actions used by SAM forces to seek opportunities for battle or avoid enemy air attacks using mobility to change operational areas."[100] These operations take three basic forms: (1) mobile ambushes [机动 设伏], which are based on an understanding of enemy activity and are undertaken in or close to areas where the enemy might operate (see next subsection); (2) mobile coverage [运动掩护], employed when forces are insufficient to cover multiple approaches to a large target area or to protect multiple targets that are located not far from one another; and (3) search and destroy operations (literally, "battlefield hunting" [战场游猎]), executed by elite forces within a relatively large battlefield seeking opportunities to strike targets.[101]

The execution of mobility operations is generally divided into several phases: preparation for the move; withdrawal from current positions; transportation of assets; occupation of new positions; and preparations for combat. While improvements in aerial reconnaissance and air attack technology make mobility more important and frequent, movement also poses challenges for the mobile element, and all moves will require the thorough study and understanding of many factors, including battlefield trends, geography, transportation, weather and moisture conditions, patterns of enemy activity, and roadway conditions. These operations also require thorough camouflage and tight command and control.[102]

Maneuver Ambush [机动设伏]. As PLAAF doctrine has placed more emphasis on mobility, initiative, and offensive spirit, its air defense writing has placed greater emphasis on maneuver ambushes by ground-based defense systems. These are operations in which SAM units shift positions and conceal themselves where enemy aircraft may be active and seek opportunities to surprise and engage enemy targets. Maneuver ambushes may either be set along enemy air routes or use various tactical measures to lure or induce enemy aircraft toward prepared

---

[100]PLAAF, 2005, p. 136.

[101]PLAAF, 2005, p. 136.

[102]PLAAF, 2005, p. 136.

ambush sites. They require thorough intelligence work and a mastery of enemy operational patterns; careful selection of ambush sites and the correct selection of either dispersed or concentrated deployment formations; strict secrecy and camouflage; and the constant or ongoing adjustment of tactical methods.[103]

Close and Quick Tactics [近快战法]. Chinese sources claim that, during the 1960s and 1970s, Chinese SAM units pioneered "close and quick" tactics. These tactics are designed to counter the enemy's ability to detect and avoid impending or actual attacks by reducing the time lag between powering on the guidance radar and missile launch. Close and quick tactics are difficult to execute, and requirements for them include (1) comprehensive analysis of the enemy's situation, tactics, and patterns of behavior; (2) strict camouflage and the control of electromagnetic emission direction, initiation time, duration, and strength; (3) mastery of the SAM system's capabilities and the ability to fully exploit (literally, "excavate" [挖掘]) its capabilities; (4) the ability to obtain air intelligence and the superior exploitation of target acquisition radars' capabilities; and (5) the ability to thoroughly understand the characteristics of the target, extent and type of electronic interference, and the optimal distance at which to power on the guidance radar.[104]

Many of these requirements depend on careful preparation prior to the engagement. The *China Air Force Encyclopedia* calls for the simplification of command and execution procedures and their thorough rehearsal prior to battle.[105] During active operations, every effort should be made to exploit the capabilities of the target acquisition radar and pass as much information as possible from it to the guidance radar, thus reducing the time during which the latter will be powered on.

Other Types of Operations

According to PLA literature, the most-common air force missions in joint campaigns (aside from purely air force operations, such as stra-

---

[103] PLAAF, 2005, p. 135.

[104] PLAAF, 2005, p. 135.

[105] PLAAF, 2005, p. 135.

tegic strikes) are seizing air superiority; providing air cover; conducting air raid and counter–air raid operations; close air support (CAS); deep air strikes; air reconnaissance; air transport; and EW.[106] Some of these missions have been discussed already, but others do not fall neatly into the "air-to-air," "air-to-surface," or "surface-to-air" categories of combat operations. Those that do not are mainly support functions: air reconnaissance, air transport, and EW. In the context of joint operations in support of the army and navy, air reconnaissance is defined as obtaining timely reconnaissance on battlefield conditions across a large area (both on the front and in the rear) and obtaining intelligence on the enemy for land and naval operations.[107] Air transport moves forces quickly, supports airborne operations, supplies logistics material, and rescues injured soldiers. Electronic countermeasures in the air and land coordinate and aim to both destroy and suppress enemy electronic installations to gain electromagnetic superiority.[108]

---

[106] Bi, 2002, pp. 140–145.

[107] Bi, 2002, p. 141.

[108] PLAAF, 2005, p. 125.

CHAPTER FIVE

Air Offensive Campaigns

Chinese military publications identify four distinct types of air force campaigns: air offensive campaigns, air defense campaigns, airborne campaigns, and air blockade campaigns. Air offensive and air defense campaigns are the most important types for the PLAAF. These can be either air force–only campaigns or, more frequently, air force–led joint campaigns that incorporate other services. These air force campaigns can also be part of broader joint campaigns, such as an island-landing campaign or joint blockade campaign.

An offensive air campaign can also be called an "air strike campaign" [空中突击战役] or "air raid campaign" [空袭战役]. It is usually part of a joint campaign, but it can also be a campaign conducted independently by air forces. Air offensive campaigns are described as large-scale, offensive air operations characterized by a high degree of initiative and autonomy relative to air defense campaigns.[1] According to Zhang Yuliang, 2006, they can be conducted to achieve either campaign-level or strategic objectives during a military conflict.[2] Zhang Yuliang also notes that various forms of air offensive campaigns exist and can be categorized according to the operational tasks and goals of a given campaign.[3] Campaign objectives are said to include seizing air superiority by destroying or weakening the enemy's aviation and ground-based air defense forces; creating proper conditions for a

---

[1] Xue, 2001, p. 371.

[2] Zhang Yuliang, 2006, p. 575.

[3] Zhang Yuliang, 2006, p. 575.

ground- or sea-based campaign by destroying or weakening the ene-my's heavy ground formations and disrupting its transportation and supply system; and striking enemy political, military, and economic targets.[4] Writings on offensive air campaigns emphasize air-to-surface strikes. Though air-to-air combat is also mentioned in writings on air offensive campaigns, air-to-air combat in an offensive air campaign is mainly discussed in the context of offensive operations, when it is dif-ficult for the PLAAF to penetrate enemy air defenses to strike enemy air assets on the ground, or defensive operations that are part of an offensive campaign, such as air interception. According to the PLAAF, the "most effective operations" are conducted over long distances, at high speed, and using intensive firepower against the enemy in strikes deep in its territory.[5]

In air offensive campaigns, as in most air operations, a great deal of emphasis is placed on surprise, camouflage, use of tactics, meticu-lous planning, and key-point strikes. Nevertheless, there is some tacit recognition that camouflage and surprise are much more difficult on the modern battlefield and that the PLAAF needs to be innovative in achieving surprise. As in publications from the mid-1990s, publica-tions of the early 2000s mention flying in from an angle (as opposed to directly toward enemy airspace) or using terrain as cover as possible tactics for evading detection, but do not consider these tactics to be adequate or reliable measures due to improvements in modern early warning detection. Overall, the tone of recent writings is cautionary: The PLAAF expects to be facing a technologically superior foe.[6]

Special Characteristics of Air Offensive Campaigns

Many official PLA writings highlight the "characteristics" [特点] of air offensive campaigns. The first is that air offensive campaigns have strong political implications, and high-level leaders decide important

---

[4]   Zhang Yuliang, 2006, p. 558.

[5]   PLAAF, 1994, p. 99.

[6]   For example, see Zhang Yuliang, 2006, p. 581.

aspects of air offensive campaigns. Often, air offensive campaigns are used to achieve national political goals—not just military goals. Often, targets are sensitive and a decision to strike them is made at the highest levels.[7]

According to PLA writings, a second characteristic is the "decisiveness" of air offensive campaigns. By grasping the operational initiative, the PLAAF can have an advantage in a campaign. An air offensive campaign, being offensive, "seizes the initiative." The campaign commander decides the scale of the operation, the forces and weapons used, the direction and timing of the operations, the targets to be attacked, and the extent of destruction. This gives the commander an advantage. Moreover, if the operational commander has time on his or her side, he or she has the opportunity to analyze the situation of the enemy and the PLAAF's situation and carefully plan a campaign.[8]

Recent writings on air offensive campaigns emphasize other characteristics that pose some challenges to the PLAAF. For example, the third characteristic of an air offensive campaign is that the situation on the battlefield is "complex," because air offensive operations take place deep in the enemy's strategic rear. Recent writings, especially, point out that this poses particular challenges to the PLAAF. Because PLAAF reconnaissance is limited and the enemy is presumed to take a variety of deceptive protective measures, it is difficult for the PLAAF to obtain accurate information on the geography, climate, air defense deployments, and operational situation in enemy territory. Therefore, it also is difficult to accurately and comprehensively predict changes in the battlefield, adding to the campaign commander's challenges when planning campaigns.[9]

Recent writings highlight another, related, area of difficulty for PLAAF air offensive campaigns: enemy air defenses. In these writings, the PLAAF appears to be referring to the air defenses of an enemy with an advanced military, which would include the United States, of course, but also, possibly, Taiwan. According to recent writings, "the enemy"

---

[7] Zhang Yuliang, 2006, p. 575.

[8] Zhang Yuliang, 2006, p. 575.

[9] Zhang Yuliang, 2006, p. 576.

has dense air defenses, and breaching its air defenses is difficult [敌对空防御严密, 突防难度大]—again because targets of an air offensive campaign are mostly located in the enemy's strategic and campaign rear. The enemy, according to these operational teaching materials, is technologically superior and has both a formidable air offensive capability and a dense air defense system. The enemy is said to already have a variety of programs for radar, air early warning aircraft, and early warning satellites, which together form a highly effective, dense reconnaissance and early warning system. Its dense firepower system uses a variety of fighter aircraft, SAMs, AAA, and a computer-centric, highly automated command-and-control system. Combined, these increase the overall effectiveness of enemy air defense operations.[10]

The difficulty of breaching the enemy's strong air defenses is emphasized in writings on air offensive campaigns. Because the PLA still does not have an extensive array of sophisticated weapons for air offense and its information systems are not advanced or integrated, it is said to be extremely difficult to successfully breach enemy air defenses that are complex and highly capable and have a strong ability to resist intrusion. Apparently, PLA strategists continue to grapple with this challenge.[11]

The final characteristic of air offensive operations is that operations are intense, consume a lot of resources, and require extensive support. Air offensive campaigns take place over a wide theater, and missions are difficult. Targets are dispersed, and there are many of them.[12]

PLAAF writings indicate that a successful offensive air campaign requires adequate preparation, camouflage and surprise, taking the initiative, striking vital points, and resolutely fighting the battle and concluding it quickly.[13] The PLA also emphasizes these guiding principles in other operations: They appear as consistent themes across many PLA campaigns and are central to PLA thought on effective combat.

---

[10] Zhang Yuliang, 2006, p. 575.

[11] See Wang Houqing and Zhang Xingye, 2000, p. 352; Zhang Yuliang, 2006, p. 576.

[12] Zhang Yuliang, 2006, p. 576.

[13] This paragraph is based on the "Requirements" section of Zhang Yuliang, 2006, pp. 577–578.

Air Offensive Campaign Objectives

An air offensive campaign can include one or more of several objectives: obtaining air superiority; destroying key enemy political, military, and economic targets; destroying the enemy's transportation and logistics supply system; and destroying the enemy's massed forces to "isolate the battlefield" and facilitate PLA ground and maritime operations. Obtaining air superiority is needed in order to conduct air strikes against targets, but the ultimate objective of an air offensive campaign is to strike political, economic, and military targets.[14]

Planning the Air Offensive Campaign

In order to carry out an air offensive campaign, PLA writings state that one needs to plan it carefully, choose appropriate timing for the campaign, and exercise the plan repeatedly. Planning and preparation includes several elements. The first element is the selection of good targets—generally, enemy "key points," such as communication or transportation hubs or other high-value targets. Targets are selected by the campaign commander and command organs under the guidance of a targeting expert. The targets, which should be adjusted as campaign conditions change, should reflect the "campaign intent" [战役

---

[14] Bi, 2002, p. 372. In a discussion of the missions of air offensive campaigns, Bi notes that, while gaining air superiority is the "primary" mission [首要任务], attacking political, military, and economic targets is the "main" mission [主要任务], and attacking logistics targets is an "important mission" [重要任务]. It is unclear whether he means that air superiority is "primary" in the sense of "first" mission, or "primary" in the sense of "first in importance." Other sources, however, indicate that air superiority is an important part of but not main objective of air offensive campaigns. Zhang Yuliang, 2006, the most recent source analyzed in this study, for example, has a somewhat different interpretation. While there is not a distinct section on "missions" or "objectives," the introduction to the chapter on "air offensive campaigns" (p. 575) briefly notes that air offensive campaigns' "missions and objectives" can be to (1) achieve air superiority, (2) weaken the enemy forces, or (3) achieve some kind of special objective. It therefore appears that the "objective" or "mission" of air offensive campaigns has evolved in recent years.

企图], [15] the target's value, enemy defenses, and PLAAF strike abilities. Second, one should concentrate one's forces toward the main objectives of the campaign and for the first strike. Third, one should strike when, where, and at a speed that the enemy least expects. Fourth, one should have a strong first strike and then follow up with subsequent strikes as needed. A great deal of emphasis is placed on a strong first strike: PLA writings emphasize that a strong first strike lays the groundwork for effective follow-up strikes and a successful campaign. [16]

Finally, defensive operations are also critical to air offensive campaigns: One should integrate defensive and offensive actions (literally, 以防助攻, or "use the defensive to assist the offensive"). In offensive air campaigns, having a strong air defense and preparing to resist an enemy's counterstrike ensure the continuity of offensive operations and an offensive posture. [17]

Operational details are based on the campaign "resolution" [决心], the forces to be used, the focus of the attack, and the campaign's operational goals. [18] With this information as a baseline, target planning includes choosing the target; locating, vetting, and analyzing information on the target; verifying its position; determining the strike parameters of the target; determining whether it is large or small in scope; evaluating and determining the order of targets to be attacked; choosing the forces that will carry out the attacks; and verifying all of the information on the target to make sure it is accurate. Having gathered targeting details, one selects a method of attacking the target(s) that will play to friendly strengths and enemy shortcomings, achieving the maximal effect quickly.

---

[15] The "operational intent" is an overall plan based on campaign goals and tasks, as well as the status of enemy air defenses, enemy targets, and PLA abilities. Determining the operational intent generally includes defining operational goals, the forces to be used, and the key targets to attack. See PLAAF, 2005, p. 100.

[16] See, for example, Zhang Yuliang, 2006, pp. 584–585. This is also consistent with the principle of "using elite forces at the beginning of the war, controlling the enemy at the first opportunity [首战使用精兵，先机制敌]" (PLAAF, 2005, p. 81).

[17] PLAAF, 2005, pp. 99–100.

[18] The "resolution" is the commander's guidance for meeting campaign objectives approved by the leadership command (Stokes, 2005, p. 258).

The next step is to establish the "campaign disposition" [战役布势]. Campaign commanders organize and deploy relevant forces—attack forces, suppression forces, cover forces, air defense forces, support forces, and campaign reserve forces—according to tasks, abilities, and specific battlefield conditions. Aviation forces are organized and arranged so as to achieve "rational" [合理] deployment in conjunction with ground forces.[19]

After determining the "campaign disposition," command elements [指挥机构] are set up and a campaign plan is established. Command elements should be able to issue uninterrupted commands in an effective, timely, and stable manner throughout the duration of the campaign.[20] Establishing a campaign plan involves examining the overall situation of the campaign; forecasting campaign scenarios that may emerge; forming plans for first strikes and follow-up attacks; ensuring safeguards for resisting enemy counterattacks; coordination; and planning for emergency situations. The campaign plan is then comprehensively assessed, tested, and verified with an eye toward its suitability for a range of operational requirements under various circumstances.[21]

The final organizational steps prior to the beginning of the campaign are to plan campaign initiation [战役展开] and operational coordination [战役协同]; coordinate "pre-combat training" [临战训练]; and plan for the support of command, logistics, and equipment. Campaign deployments must reflect the importance of speed, timeliness, and concealment. Operational coordination is often achieved by organizing forces around the air units responsible for carrying out the central tasks of the operation.[22]

---

[19] PLAAF, 2005, p. 100.

[20] PLAAF, 2005, p. 100.

[21] PLAAF, 2005, p. 100.

[22] PLAAF, 2005, p. 100.

Strike Methods and Force Composition

As discussed in Chapter Four, the terms *air strike* [空中突击] and *air raid* [空袭] are sometimes used interchangeably. Both refer to air-to-surface strikes. An air strike tends to be a fairly focused mission using a single aircraft or combat group against a single target or target group. An air raid is generally broader in scope and seeks to achieve broader objectives. Air strikes or raids are the most common form of attack in an air offensive campaign, though air-to-air operations are also used. In some cases, the term *air offensive campaign* is actually synonymous with a series of air strikes or air raids.

PLA sources generally agree on three types of commonly used methods for striking surface targets from the air.[23] These include concentrated strike [集中突击], simultaneous strike [同时突击], and continuous strike [连续突击].

Concentrated strike uses a large number of forces in a limited area within a short period of time, either from multiple directions or from a single direction, in order to shock the enemy and completely destroy targets. It is generally used against robust targets and can be useful for first strikes against important targets or to make use of surprise.[24]

Simultaneous strike uses several formations against a target system or several targets, in order to paralyze the enemy. The main strike group operates in waves or simultaneous strikes against one area or target, or several targets in different areas. It is mainly used against enemy industry, transportation hubs, or logistics, petroleum, airbases, or other facilities. This kind of strike can also disperse enemy air defense forces and firepower or blockade a region, obstructing enemy movements. It is used only when it is really needed or when one has absolute force

---

[23]   Bi, 2002, pp. 374, 377; PLAAF, 2005, p. 72. Referred to as "air strikes" [突击] in most sources but also referred to as "air raids" [空袭] in many other sources, including under the entry on "air raid" in the *China Air Force Encyclopedia* (PLAAF, 2005, p. 72). Most of the information in this section is derived from Bi, 2002, pp. 374, 377, and from PLAAF, 2005, p. 72. See also PLAAF, 1994, pp. 168–169, for an older description of methods of strike.

[24]   See also PLAAF, 1994, pp. 168–169, for an older description.

superiority because targets are scattered and it requires a large number of forces and a lot of coordination.[25]

Continuous strike consists of strikes over a relatively long period of time, in order to suppress, harass, and exhaust the enemy. Continuous strikes also can prevent the enemy from recovering from an earlier strike. They often occur just after or at the same time as a concentrated attack; forces used often depend on battle damage assessment results from previous strikes. When consecutively striking, according to some sources, strategic missile forces [战略导弹部队] and missile submarine forces [导弹潜艇部队] may be used to strike targets with strong air defenses, robust targets, or targets spread over a large area. These strikes will be followed by air force strikes against smaller targets or targets with weaker air defenses. Because they occur over a long period of time, continuous strikes are difficult to camouflage or use surprise and are difficult to sustain. Changing tactics frequently prevents the enemy from taking advantage of predictable patterns of operations.[26]

Organization and Deployment of Forces in an Air Offensive Campaign

There are several types of combat groups that are involved in air offensive campaigns: the strike group [突击集群], the suppression group [压制集群], the cover group [掩护集群], the support group [保障集群], the air defense group [防空集群], and the operational reserve

---

[25] See also PLAAF, 2005, pp. 123–124, for definitions of different types of air strikes.

[26] Most of the information in this paragraph is derived from Bi, 2002, pp. 374, 377, and from PLAAF, 2005, p. 72. Note that *strategic missile forces* generally refers to nuclear missile forces, but it is not clear whether that meaning is intended in this context or whether it is intended to simply refer to the Second Artillery Force, which operates both nuclear and conventional missiles. PLAAF, 1994, p. 169, mentions the need for changing tactics so the enemy does not detect a pattern of operations. Other, more-recent sources also mention the need to "flexibly" apply different tactics—e.g., PLAAF, 2005, p. 41, in two separate entries on "integrated operations in the air" and "air superiority." More-recent sources appear to be more concerned with flexible tactics to suit the operational environment, however, than with preventing the enemy from detecting a pattern.

[战役预备队].[27] In most cases, forces are organized around the strike group in offensive air campaigns. The forces deployed would depend mainly on the type of air offensive campaign, and, in many cases, not all of them would be used. Depending on the size of the campaign, however, at least elements of the strike group, suppression group, cover group, and support group would likely be used in most air offensive campaigns.

The strike group is the group that carries out the air raid and attack in an air offensive campaign. It is composed mainly of bombers [轰炸机], fighter-bombers [歼击轰炸机], and attack aircraft [强机] forces—the main body of offensive air campaign operational strength (support forces and paratroops can also be added to the group as needed). Its mission is to "smash, destroy, and eliminate" target systems with the cooperation of the suppression, cover, and support groups. Whether the group is large or small depends on the target, enemy situation, campaign objectives, and the enemy's threat to one's forces. To effectively destroy or eliminate the target, usually the forces will be divided into attack waves [突击波], which are further divided into echelons [梯队].

The suppression group is responsible for attacking enemy air defense systems in order to ensure that the strike group can carry out its attack smoothly. Usually, this group comprises bombers, fighter-bombers, attack, and electronic countermeasure [电子对抗] forces.[28] Its mission is to carry out electronic and firepower suppression, destruction, and blockades, and to open penetration corridors [突防走廊] for other groups (such as the strike group). This group can carry out the first strike as well as conduct follow-up strikes. The size of the suppres-

---

[27] This characterization of the types of forces used in an air offensive campaign is based on Bi, 2002, pp. 377–378, and PLAAF, 2005, pp. 100–101. Bi also mentions an electronic countermeasure group [电子对抗集群], but no reference to an electronic countermeasure group can be found in PLAAF, 2005. Instead, PLAAF, 2005 refers to "electronic countermeasures forces" [电子对抗兵] as being part of the suppression group.

[28] The description of the suppression group is based on PLAAF, 2005, pp. 100–101. It is worth noting that the PLAAF, 2005, description of the suppression group is somewhat different from the slightly older description in Bi, 2002, p. 377, which does not mention EW forces.

sion force depends mainly on the density, deployment, and operational capabilities of the enemy's air defense weapons and the objectives of the PLAAF campaign.[29]

The cover group consists of the forces that are responsible for protecting the attack, suppression, and support groups' operations. It usually consists of fighters and long-range SAM forces. Its missions include eliminating, suppressing, or driving away enemy fighters and early warning command aircraft, to ensure that the strike, suppression, and support groups do not face much resistance from enemy airpower. The scale of cover group operations depends on how effective its coverage is, the degree of enemy threat, and the scale of the aircraft to be covered.[30] Figure 5.1, drawn from the *China Air Force Encyclopedia*, illustrates the physical relationship between the strike group, suppression group (including an electronic countermeasure group), and cover group. How the PLA intends to coordinate the operations of the cover group and the strike group is not completely clear from the sources analyzed for this study. One possible approach to this problem in a Taiwan scenario that is at least consistent with PLA publications is explored in Chapter Ten.

The support group provides logistic, intelligence, and other support to the strike, suppression, and cover groups so that they can smoothly achieve their air offensive campaign objectives. Usually, the support group includes reconnaissance aviation [侦察航空兵], early warning and command [预警指挥], aerial refueling [空中加油], aerial rescue [空中救护], meteorological [气象部队] units and radar, communications, and technical surveillance forces.[31]

The air defense group consists of forces to carry out the counterstrike aspect of an air offensive campaign.[32] They include fighters, SAMs, AAA, and other air defense forces. Their mission is to defend

---

[29] The suppression group is a new element, indicating an adjustment since 1994 in how forces would be deployed. In PLAAF, 1994, responsibility for suppression fell under the "cover group." There is no mention of a separate suppression group.

[30] PLAAF, 2005, p. 101.

[31] PLAAF, 2005, p. 101.

[32] The air defense group definition is based on Bi, 2002, p. 378.

Figure 5.1
Penetrating Enemy Defenses

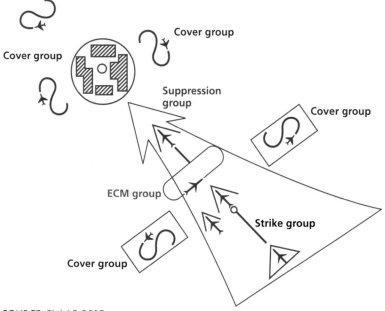

SOURCE: PLAAF, 2005.
NOTE: ECM = electronic countermeasure.
RAND *MG915-5.1*

against enemy attack on airfields, command headquarters, and other critical targets.

The operational reserve consists of some fighters, bombers, fighter-bombers, and attack aircraft, as well as some airborne forces.[33] Its main mission is to strengthen forces where needed in order to speed up the progress of the campaign, exploit combat results, or complement forces at a critical moment.

---

[33] Reserve units are not mentioned as a separate entry in PLAAF, 2005, but almost certainly exist. They do appear in PLAAF, 1994, and in Bi, 2002, p. 378.

Deployments on the Ground and in the Air

The groups discussed in the previous section may not be deployed together on the ground. During a mission in a campaign, forces are deployed in the following order: First, command, logistics, and technical support are deployed covertly. Fighters, SAM forces, AAA, and electronic countermeasure forces are deployed next. "Offensive air forces" (this probably refers to the strike group) are deployed last.

Forces in an air offensive campaign are deployed both on the ground and in the air. Ground deployments are divided into front line, second line, rear area, and "other deployment." On the front lines are fighters that have air defense, cover, and some strike [以歼代强] duties. The second line includes attack aircraft and some fighter-bombers for suppression duties. Forces deployed in the rear include bombers that can fly long distances, the operational reserve, airborne early warning and command aircraft, and aerial refueling aircraft. Electronic warfare aircraft and other support forces are distributed as appropriate.[34]

Air deployment is accomplished either as mixed aircraft including cover forces or as small detachments of several strike groups with no cover. The former usually consists of a multifunctional group in which the strike group forms the core. Electronic countermeasure and suppression forces deploy first. Cover forces surround and protect those forces, and the early warning and command aircraft acts as the core of the deployment.[35]

Carrying Out the Campaign

An offensive air campaign consists of a series of actions, most likely largely sequential: conducting information operations, penetrating enemy defenses, conducting air strikes, and protecting against counterstrikes. The fourth task would be implemented throughout the campaign, possibly even beginning before the overall campaign began, so as

[34] Bi, 2002, p. 378.

[35] Bi, 2002, pp. 378–379.

to counter enemy preemptive strikes aimed at thwarting the air offensive campaign. The other stages are generally but not strictly sequential (and are not explicitly described as such). For example, while the bulk of "conducting information operations" is a preparatory action that occurs before the campaign, some aspects of "conducting information operations" would be carried out throughout the campaign.[36]

Action One: Conduct Information Operations

In its most recent publications, the PLA identifies the first action of an air offensive campaign as "conducting information operations" [组织信息作战]. This action is first noted in the 2006 edition of *Study of Campaigns*.[37] It is consistent with the PLA's increased emphasis on "informationized warfare"—which focuses on the importance of information and data links to modern weapon systems and command and control, reconnaissance, and information operations in modern warfare. Previous descriptions of air offensive campaigns have not included a separate task that includes all of the elements described in the "Conduct Information Operations" action mentioned in Zhang Yuliang, 2006.[38] The steps required within this phase are said to be: conducting information reconnaissance, conducting an information offensive against the enemy (electronic interference and deception, firepower destruction, and computer network attacks), and conducting information defense against the enemy.[39]

Conduct Information Reconnaissance [组织信息侦察]. Conducting information reconnaissance consists of using space, air, naval, and ground reconnaissance assets to gather intelligence on enemy

---

[36] See PLAAF, 2005, pp. 99–100; Zhang Yuliang, 2006, pp. 579–588.

[37] Zhang Yuliang, 2006, pp. 579–581.

[38] Zhang Yuliang, 2006. Neither PLAAF, 2005, pp. 99–100, nor Wang Houqing and Zhang Xingye, 2000, includes a description of "conducting information operations." For example, Wang Houqing and Zhang Xingye, 2000, begins its discussion of actions to take during an air offensive campaign with carefully selecting targets and finalizing air deployments, rather than conducting information reconnaissance and other steps laid out in the 2006 edition of *Study of Campaigns*. See Wang Houqing and Zhang Xingye, 2000, p. 354, and Zhang Yuliang, 2006, p. 579.

[39] Zhang Yuliang, 2006, pp. 579–581.

information operations, including the following: enemy methods for information operations, information command hubs, computer network nodes, and the quality and functionality of information warfare equipment. Reconnaissance is completed before offensive operations begin. Tasks focus on obtaining targeting data on the enemy's early warning system, air command-and-control system, surface-to-air air defense, and command and guidance systems, and mechanized firepower command system. By gathering this information in advance, the PLAAF hopes to attain an advantage for strikes at the beginning of the war.

Conduct an Information Offensive [组织信息进攻]. This uses "integrated 'soft' and 'hard' attack actions" ["软", "硬"一体的攻击行动] against the enemy's information system to create conditions for achieving information superiority. Soft actions include electronic interference and deception. Hard actions include the destruction of targets using bombs, missiles, and other firepower.[40]

Electronic Interference and Deception [电子干扰, 欺骗]. According to PLA operating concepts, the PLA, which apparently views its ability to conduct electronic interference as "limited," carries out electronic interference at the same time as air offensive operations.[41] Electronic interference is concentrated in the main direction of the campaign during crucial times, such as the takeoff of strike groups, during operations to penetrate air defenses, and during the strike itself. During these times, the PLA will use intense electronic interference to suppress enemy electromagnetic targets and weaken the enemy's ability to conduct information operations [信息作战能力]. The PLA will attack enemy electromagnetic targets to weaken the enemy's ability to conduct information operations, focusing interference on enemy reconnaissance and early warning satellites, airborne early warning and command aircraft, ground-based long-range early warning radars, the

---

[40] Zhang Yuliang, 2006, p. 580.

[41] Zhang Yuliang, 2006, p. 580. The text reads, "In a situation in which our electronic interference strength and measures are limited, electronic interference actions usually are carried out in coordination with air offensive actions" [在我电子干扰力量和手段有限的情况下, 电子干扰行动通常与空中进攻行动结合进行。].

radars of intercepting aircraft, SAM guidance radars, and command guidance systems. Furthermore, teaching materials admonish soldiers carrying out information operations to use electronic deception, lure the enemy, and create illusions that make it difficult for the enemy to distinguish between what is real and false.[42]

Firepower Destruction [火力摧毁]. In addition to soft strikes on the enemy's air defense system, the PLAAF also advises carrying out hard strikes, or "firepower destruction," to cripple or blind the enemy's air defense system. The principal approaches for conducting "firepower destruction" strikes include using antiradiation unmanned aerial vehicles (UAVs), antiradiation ballistic missiles, antiradiation cruise missiles, or air-launched antiradiation missiles to attack important enemy early warning radars, missile-guidance radars, and other types of electromagnetic targets.[43]

Before strike formations [突击编队] take off, a portion of forces that are good at penetrating defenses will attack certain enemy command-and-control centers, reconnaissance and early warning systems, and air defense system nodes [节点] in order to open up gaps in the enemy's air defense system. These breaches will undermine the enemy's ability to organize effective interception operations and open up a path for subsequent strikes against enemy air defenses.[44]

Computer Network Attack [计算机网络攻击]. In this action, the PLAAF will use computer network forces to attack enemy computer networks (stealing from, changing, erasing, deceiving, or obstructing the networks), in an attempt to paralyze or weaken the enemy's air defense operational capabilities and create conditions for air strike operations.[45]

Conduct Information Defense [组织信息防御]. While conducting an information attack, the commander and command organs are expected to conduct an "information defense"—to defend against

---

[42] Zhang Yuliang, 2006, p. 580.

[43] Zhang Yuliang, 2006, p. 580.

[44] Zhang Yuliang, 2006, p. 580.

[45] Zhang Yuliang, 2006, p. 580.

attacks on the PLA's own information systems. Doing so consists of defending against enemy reconnaissance, countering electronic interference, resisting the enemy's hard destruction of the PLA's information systems, and defending against enemy network attacks on the PLA.[46]

Defend Against Enemy Information Reconnaissance [防敌信息侦察]. The commander and command organs should focus on concealing the campaign intent and key objectives and on conducting actions to resist enemy information reconnaissance. To decrease the effectiveness of enemy reconnaissance, the PLA will use a variety of methods, including concealment, dispersal of forces, camouflage, the use of decoys, disinformation, feints, information security, and the control and protection of news media, the electromagnetic spectrum, and computer networks.[47]

Counter Electronic Interference [反电子干扰]. Countering electronic interference is central to conducting information defense. It means focusing on radar counterinterference, using technical means to defeat enemy electronic interference and ensure that PLA information systems operate normally.[48]

Resist Enemy Firepower Destruction [抗敌活力摧毁]. The PLA plans to use such measures as emission control, frequently switching transmitting stations, mobility and evasion, and firepower cover to avoid or minimize destruction from enemy attack on PLA information systems.[49]

Defend Against Enemy Network Attacks [防敌网络攻击]. Finally, the PLA is concerned about attacks on its computer networks. To avoid intrusion into its networks, writings exhort commanders to manage computer networks (especially the network host) carefully;

---

[46] This section is based primarily on the most-recent PLA publication on air offensive campaigns analyzed for this study, Zhang Yuliang, 2006, pp. 580–581. However, Zhang Yuliang builds on and is consistent with previous writings on air offensive campaigns, including PLAAF, 2005, pp. 99–100; Bi, 2002, pp. 371–384; and Wang Houqing and Zhang Xingye, 2000, pp. 354–356.

[47] Zhang Yuliang, 2006, p. 581.

[48] Zhang Yuliang, 2006, p. 581.

[49] Zhang Yuliang, 2006, p. 581.

restrict administrative rights at all levels; carefully guard passwords, verbal orders, and addresses against theft; and isolate computers to make every effort to reduce or prevent enemy outsiders from hacking into the network and destroying it using either physical means or viruses.[50]

Action Two: Penetrate Enemy Defenses

This is an operation to break through enemy air defense systems (presumably in a more comprehensive way than required for the firepower destruction element of the information offensive described in the previous section).[51] It is conducted by a strike group [集群] or formation [编队]. According to PLA writings, breaking through the enemy air defenses is a prerequisite for conducting an air offensive campaign and can have a great impact on the success (or failure) of the campaign. PLA concern about the difficulties of penetrating enemy defenses, especially those of the United States and Taiwan (this is strongly implied, though not explicitly stated), also is evident:

> right now, our main operational opponent has already built a dense long-, medium-, and short-range, and high-, medium-, and low-altitude air defense system, but we have limited numbers of highly capable offensive fighters, and our overall ability to penetrate defenses is weak, so it is very difficult to penetrate the dense resistance from enemy aircraft, missiles, and guns. Therefore, it is necessary to take all kinds of flexibly applied methods and measures to ensure strike forces successfully penetrate enemy ground-based air defenses.[52]

---

[50] Zhang Yuliang, 2006, p. 581.

[51] This section is based on Zhang Yuliang, 2006, pp. 581–584, except where otherwise noted, but there also is a much more general and concise but consistent section on penetrating air defenses in PLAAF, 2005, pp. 122–123.

[52] Zhang Yuliang, 2006, p. 581. The Chinese is
目前, 我主要作战对象已建立了远中近程, 高中低空相结合的严密防空配系, 而我高性能的进攻性飞机数量有限, 总体突防能力不强, 要突破敌机弹炮的密集抗击, 难度较大。因此, 必须灵活运用多种突防方式和手段, 以确保突击兵力顺利突破地对空防御。

The PLAAF's approach to penetrating enemy air defenses is to seek asymmetrical methods for use in the attack, including concealment tactics—discussed in detail later in this section—as well as suppression and direct strikes against the enemy's air defenses. Discussion of strikes against the enemy's air defenses emphasizes achieving partial air superiority, opening air corridors, and limiting the amount of direct engagement with the enemy.[53]

According to the *China Air Force Encyclopedia*, height and speed of penetration will depend on the capabilities of the enemy's aircraft and air defenses, but "high-speed entry at a low altitude is often the most effective method."[54] It is notable that the PLAAF considers flying at low altitude to be an effective tactic, either in operations to penetrate enemy defenses or in attack operations. Flying at low altitude is a tactic that has fallen out of favor in most Western militaries for a number of reasons, including proliferation of short-range air defense systems. In a Taiwan scenario, it might make some sense because the aircraft would not be at risk for being shot at when flying at low altitude over water (the Taiwan Strait), might succeed in flying under the engagement envelopes of Patriot missiles at low altitude, and could avoid detection over Taiwan by flying in its valleys or plains below the tall mountains running down its center, and quickly scaling the mountains and hugging the terrain again on the other side if necessary. Nonetheless, given Taiwan's mountainous, uneven terrain, and densely populated west coast, flying at low altitude could also be a risky tactic.

This paragraph is interesting for a couple of reasons beyond the discussion of air defense. First, it demonstrates that the PLAAF is very concerned about its own capabilities in the face of an operationally superior opponent. Second, it advances an argument for using asymmetric capabilities to counter that superior foe. Interestingly, neither of these concepts appeared in Wang Houqing and Zhang Xingye, 2000, but did appear in a much earlier work on PLAAF tactics (PLAAF, 1994, pp. 93, 117, 121–123, 153–154, 164). The PLA seems to be revisiting issues of the asymmetry of capabilities.

[53] See Zhang Yuliang, 2006, pp. 581–583. For example, in discussion of meeting the enemy, the text suggests retreating when the enemy counterattacks or when enemy fighters intercept air forces attempting to penetrate enemy offenses (last paragraph of p. 583).

[54] PLAAF, 2005, pp. 122–123.

PLA publications do not exude confidence that the PLA will be able to hold comprehensive air superiority to allow freedom of maneuver. Rather, these writings emphasize using measures that will allow PLA forces to achieve operational objectives in the face of an operationally superior opponent. A theme running throughout PLA writings is evasion and deception of the opponent. It is one of the two major methods discussed for penetrating enemy air defenses. The other method is suppression of enemy air defense (SEAD).

Hide the Real and Show the False; Conceal the Attack Against the Defenses [隐真示假，隐蔽突防]. This is one of the two major methods discussed for penetrating enemy air defenses. This method entails PLA aviation strike formations using a variety of measures to "hide the real and show the false, to lure the enemy, and secretive actions to take them by surprise and attack their air defense system."[55] With fewer expected losses, higher rates of success, and fewer forces needed, this approach is said to yield twice the results with only half the effort if performed successfully.[56] However, according to PLA writings, it will be more difficult to pull off in the future as the enemy achieves a more transparent battlefield with an advanced intelligence and early warning system that has a comprehensive, around-the-clock reconnaissance capability. To achieve surprise attack in this environment will require much more-innovative means of achieving surprise. A commander flexibly combines the following measures according to the operational realities of the campaign:

1. Choose concealed air routes for attack against defenses [选择 隐蔽的航线突防]. For example, PLA sources suggest choosing routes that avoid or delay discovery by radar.[57]
2. Choose advantageous routes so as to slice through the enemy defenses at an angle [采取有利的航行剖面突防] (as opposed

---

[55] Zhang Yuliang, 2006, p. 581.

[56] Zhang Yuliang, 2006, pp. 581–582.

[57] Zhang Yuliang, 2006; PLAAF, 2005, pp. 122–123.

to flying directly toward the defenses). Minimizing flight within enemy air defenses is also recommended.[58]

3. Choose to attack when the weather conditions are advantageous to the offensive.[59]

4. Use deceptive measures during the attack.[60] For example, the PLAAF recommends that its forces form columns with a distance between them that is smaller than the enemy's ability to resolve individual targets before they enter enemy fighter air interception lines or approach enemy air defenses. Once the air combat forces enter the enemy fighter air interdiction lines or engagement range of ground-based air defenses, the distance between columns should exceed the kill radius of enemy SAMs. Follow-up formations should be close enough together that they can pass through enemy air defenses in less time than it takes for enemy ground-based air defenses to complete their engagement cycles.[61]

5. Use airborne electronic equipment, stealth technology, and tactics to increase the strike formations' ability to penetrate enemy defenses.[62]

Using "Soft" Suppression and "Hard" Destruction, Storm Through the Defenses ['软' 压, '硬' 毁, 强攻突防]. A strong attack on enemy defenses integrates "soft suppression" and "hard destruction" to breach enemy air defense systems, open air defense penetration corridors [突防走廊], support strike forces' smooth passage through the region covered by enemy air defenses, and carry out strikes against prepared targets. The advantage of carrying out a strong attack against enemy defenses is that one has greater control over the timing and location of the attack, and strikes can directly destroy the enemy's

---

[58] Zhang Yuliang, 2006, p. 582; PLAAF, 2005, pp. 122–123.

[59] Zhang Yuliang, 2006, p. 582.

[60] Zhang Yuliang, 2006, p. 582.

[61] PLAAF, 2005, pp. 122–123.

[62] Zhang Yuliang, 2006, p. 582.

remaining strength, break its air defense system, and facilitate follow-on strikes.

The strike group's composition depends on the scale of the strike forces and position of the target; it can include formations for reconnaissance, electronic countermeasures, suppression, cover, strike, and support.[63]

Conduct Air Reconnaissance [组织空中侦察]. Reconnaissance forces conduct detailed reconnaissance against the target, examining enemy air defense deployments and the precise position of targets, to facilitate follow-on electronic interference and strikes.[64]

Conduct Electronic Interference and Suppression [组织电子干扰压制]. Forces conducting electronic countermeasures should focus on the use of electronic interference, adopting both "area" (standoff) jamming and more-targeted, escorted jamming to suppress enemy electromagnetic targets. At the same time as "soft" jamming, suppression forces should use antiradiation weapons to strike enemy early warning radars, air defense missile guidance radars, and other electronic targets. This will open one or more bands of strong electronic interference in the enemy's air defense network, blinding enemy radars, cutting off communications, making air defense weapons lose effectiveness, complicating command, and lowering enemy air defense systems' combat effectiveness.[65]

Conduct Firepower Suppression [组织火力压制]. Following electronic suppression, firepower suppression forces concentrate their firepower and destroy enemy SAMs and air defense bases in the area of the strike forces' air routes, in order to create air corridors that are free of threat from ground-based firepower. Suppression forces also can suppress and blockade enemy airfields basing the fighters that pose the greatest threat. Land, naval, and Second Artillery forces also can play a role if they are part of the campaign: Ground-based or shipborne artillery can be responsible for suppression of targets that are within their

---

[63] Zhang Yuliang, 2006, pp. 582–583.

[64] Zhang Yuliang, 2006, p. 583.

[65] Zhang Yuliang, 2006, p. 583.

range. Second Artillery forces can suppress enemy targets, such as airfields and important air defense missile bases.[66]

Conduct Strike Forces' Penetration of Enemy Defenses [组织突击兵力突防]. With electronic and firepower suppression completed, the strike forces are, theoretically, in a better position to conduct their attack. Because the strike forces carry out the main mission, their protection is critical to successfully accomplishing the mission. They should fly along the air corridors that the interference and suppression forces have opened up. Using appropriate formations, they should pass quickly through the enemy's air defense firepower zone. When they run into obstruction by surviving enemy air defense power, they should use maneuvering methods to evade them—going to great lengths to avoid engaging with enemy combat aircraft if they are intercepted. With the protection of the cover forces, they should rapidly leave the scene, flying toward the target to be attacked.[67]

Conduct Air Cover [组织空中掩护]. Air cover formations use a combination of area and escort cover to destroy or expel enemy combat aircraft that pose a relatively great threat to the campaign's strike formation. According to PLA documents, when the enemy's intercepting forces do not pose a great threat to the strike formation, then the cover forces should avoid engagement; when they are a great threat to the strike formation, however, then the cover forces should engage the enemy in the air to destroy or expel the enemy intercepting forces. Even under these circumstances, however, the air campaign's cover forces should avoid excessive engagement with enemy aircraft to prevent the strike formation from losing cover for an extended period.[68]

Action Three: Conduct Air Strikes

The main objective of the campaign—striking enemy targets—is carried out during the third phase; the first two lay the groundwork for these attacks. PLA publications note that, with the development of more-sophisticated airborne weaponry and platforms, "traditional"

---

[66] Zhang Yuliang, 2006, p. 583.

[67] Zhang Yuliang, 2006, p. 583.

[68] Zhang Yuliang, 2006, p. 584.

aerial attacks—single-service air strikes and bombing runs limited to one type of aircraft—will be replaced by air offensive campaigns consisting of multiservice attacks that incorporate multiple types of aircraft, long-distance raids, and attacks outside the defensive zone. At the same time, air strike capabilities are also becoming stronger, and it is possible to employ surgical strikes against the enemy with laser-guided precision weapons or use antiradiation missiles to strike an enemy area. Other bombs, such as electromagnetic pulse bombs [电磁脉冲弹] and carbon filament bombs [石墨弹], can cause the enemy to lose its ability to function.[69]

The PLAAF divides an air strike into two subcategories: the initial strike and follow-on strikes.

The Initial Strike [首次突击]. The initial strike commences the strike phase of the air offensive campaign. The goal is to weaken the enemy's campaign operational ability, paralyze its operational system, and facilitate follow-on strikes. In informationized air raid operations, the first battle lays the foundation for the rest of the war and can even decide the outcome. For example, U.S. air raids against Libya and Israeli air raids against Iraq's nuclear facilities are both said to have achieved war aims within one strike.[70]

Because the initial strike is a key part of the strike plan, the PLA plans to use the best forces and equipment and most (approximately 80 percent) of its offensive assets, as well as support from the army, navy, and Second Artillery.[71] During this initial strike, PLA writings state that one should adhere to three guidelines:[72]

1. Concentrate the force to strike vital targets. Most PLA sources list the targets in the following order: combat airfields and air-

---

[69] Zhang Yuliang, 2006, p. 584.

[70] Zhang Yuliang, 2006, p. 584.

[71] Bi, 2002, p. 383. Zhang Yuliang, 2006, pp. 584–586, also discusses initial and subsequent strike. The 80-percent figure presumably means 80 percent of the forces allocated to the air offensive campaign, not 80 percent of the entire national inventory.

[72] Bi, 2002, and Zhang Yuliang, 2006, both outline these three basic guidelines, though not in the same order or at the same level of detail.

craft, early warning radar stations, SAM and AAA installations, command and control centers, and communications hubs.[73] In terms of tactics, the PLAAF plans to achieve battlefield air superiority by adopting a combination of long-range standoff attacks [防区外远距攻击], overhead bombing [临空轰炸], carpet bombing, and precision strikes. Strikes can be multidirectional, concentrated, or continuous, using multiple waves to achieve tactical superiority.[74] If the scale of the attack forces is relatively large, the PLA also may attack enemy political, economic, and cultural centers, important resources, water and electric installations, and other targets that affect the enemy's wartime potential, as well as military bases and installations. The forces and weapons used will depend in part on the types of targets attacked and the battlefield environment. To attack targets in the enemy's strategic rear or targets that have dense air defenses, the PLA advises using mainly air-launched long-range PGMs to attack targets from outside of the enemy's defensive zone, or using stealth aircraft to carry out a stealthy attack. On the other hand, the strike formation can attack targets with relatively weak air defenses near or somewhat near the enemy's defensive zone under the cover of strong electronic suppression.[75]

2. Intensify force rotations and increase the number of available runways, take-off points, and refueling stations. In order to ensure that the PLA is able to employ maximum force on its initial strike, it might take over airfields or use airports usually allocated for civilian use, reserve airfields, decommissioned airfields, and old airports to increase the number of airfields available for taking off and landing PLA aircraft. The PLA might also begin rotating deployments inside and outside of the battle

---

[73] Bi, 2002, p. 383.

[74] Zhang Yuliang, 2006, p. 585.

[75] Zhang Yuliang, 2006, p. 585.

zone to increase the density of strikes and increase the number of available airfields.[76]

3.   Closely coordinate all airborne strike and support actions under an early warning and command aircraft. Under the early warning and command aircraft, all air formations should work together to strike planned targets. Electronic interference formations should jam and suppress all types of enemy radar, command, and communication installations. Suppression forces should suppress and destroy the enemy's ground-based air defense weapons and enemy fighter bases posing the greatest threat, in order to clear obstacles for the strike formations to successfully conduct their air strikes. Cover formations should position themselves to conduct air patrols in the direction from which the enemy is most likely to fly if it conducts an air raid, and destroy the enemy coming to strike. After all of the strike formations have completed their strike missions, they should quickly return to their air bases, minimizing their impact on the ability of follow-on strike formations to maneuver.[77]

Follow-On Strikes [尔后突击]. Follow-on strikes are strike actions that are carried out after the initial strike. The goal of these is to destroy enemy targets that have not been destroyed yet or to strike targets that have not yet been struck or that have reappeared. Doing this solidifies or expands the results of the initial strike to reach overall operational goals. Battle damage assessment (BDA), continuous attacks to exploit results of initial attacks and prevent the enemy's recovery, flexible combat approaches, and sustainment of the PLA's combat ability over a long period of time are all necessary steps for achieving these goals.[78] PLA writings advise that planners of subsequent strikes should quickly assess initial strike results (i.e., do BDA) and take full advantage of the

---

[76] Zhang Yuliang, 2006, p. 585.

[77] Zhang Yuliang, 2006, p. 585.

[78] This is from Bi, 2002, and Zhang Yuliang, 2006, both of which have significant discussions of follow-on strikes (Zhang Yuliang, 2006, is more detailed, but Bi provides a comprehensive overview).

results of the first strike. The campaign command should use a variety of means, such as space reconnaissance, air reconnaissance, sea-based reconnaissance, ground-based reconnaissance, and "wireless technology reconnaissance" [无线电技术侦察] (possibly a reference to electronic intelligence collection) to quickly assess the results of the first strike and changes in the enemy's force deployments, and then, on the basis of this, adjust the intentions for the follow-on strikes and organize forces in preparation for further deployments.[79]

Flexible combat approaches could include relatively large-scale concentrated attacks, as well as smaller-scale air strikes and special combat operations, such as airborne landings.[80] Follow-on strikes should take full advantage of the results of the initial strike, and every effort should be made to limit the time between the first strike and any follow-on strikes. If follow-on strikes occur soon after the initial strike, the enemy will not have time to recover from the shock of the initial strike. The PLAAF (and other forces as relevant) should concentrate its forces and weapons to carry out a renewed strike on important enemy targets and concentrated and continuous strikes to maintain the pressure of continuous, nonstop strikes against the enemy, directly achieving planned objectives.[81]

PLA publications state that priority should usually be given to strikes against important targets that have suffered partial damage so as to eliminate them completely. Forces should then be concentrated for follow-on strikes against lower-priority targets and newly discovered targets. Some of those follow-on strikes will occur against political targets, such as government organs and radio and television stations. Others will include economic targets, such as transportation hubs, energy resources, water and electric installations, and military targets, such as command centers, air and naval bases, missile emplacements, and logistics facilities. Target selection should be flexibly determined based on the overall strategic needs of the campaign and PLAAF strike capabilities. Follow-on strikes are usually sustained for a relatively

---

[79] Bi, 2002, p. 384; Zhang Yuliang, 2006, p. 586.

[80] Bi, 2002, p. 384, provides this detail on flexible combat approaches for follow-on strikes.

[81] Bi, 2002, p. 384; Zhang Yuliang, 2006, p. 586.

long period of time, so forces and weapons should be used carefully to ensure continued strike capability.

Action Four: Resist Enemy Air Counterattacks

This is an ongoing defensive operation that lasts throughout the offensive air campaign. The operational objective is to ensure that, during offensive campaigns, deployments remain stable, and critical Chinese facilities (such as early warning radars, air and missile bases, and leadership compounds) remain safe, and to facilitate a successful air offensive campaign. There are two types of operations to resist enemy counterattacks. In the first, the PLA guards against an opponent that is attacking or preparing to attack preemptively to destroy PLA campaign preparations. In the second, the PLA protects its returning (or returned) forces against enemy attack. Airfields, command-and-control centers, communication hubs, and other critical targets are all considered likely targets of enemy attack. The campaign commanders are tasked with resisting counterstrike operations.

Resistance Operations [抗击作战]. For resistance operations, forces are deployed early to defend against preemptive strikes. Deployments encircle important airports, command installations, conventional guided-missile bases, and early warning systems. Resistance can include the following actions.

Air Intercepts [空中拦截]. For air intercepts, the PLA writes that it would allocate a portion of fighters in the direction of the enemy's most probable flight path. The fighters patrol in the air; when the enemy arrives, they intercept the enemy at the furthest point possible from the intended target. Fighters awaiting battle on the airfield will take advantage of this interception at a distance to mobilize in echelons and carry out continuous interception against the enemy coming to strike in order to destroy most of the enemy forces in the air, disrupt the enemy's strike deployment, and facilitate resistance from the ground.[82]

Ground Intercepts [地面拦截]. In a ground intercept, the PLA uses a portion of its medium- and long-range surface-to-air guided

---

[82] Zhang Yuliang, 2006, p. 587.

missiles deployed across the front of the defensive zone. As in the air intercept, the objective is to intercept the enemy at as great a distance as possible. Ground intercepts are directed at aircraft at a variety of ranges (short to long) and altitudes (low, medium, and high). The ground-based missiles attempt to implement lines or rings of interception attacks against approaching enemy aircraft, eliminate enemy aircraft before they drop their bombs, and intercept cruise missiles before they reach intended targets.[83]

Protective Actions [防护行动]. Protective actions are passive defense measures that are taken to avoid enemy air raids or to minimize their impact should they occur. "Effective protection" [有效的防护] is also considered to be essential to conduct air operations smoothly and to ensure the stability of deployments. PLA writings point out that, in an "informationized" war, battlefields are transparent, air weaponry is destructive, and defensive missions are difficult. Therefore, defenses need to be well organized and tightly controlled under a unified organization encompassing all services and forces, armed police forces, and mobilized civilians. Protective actions can include fortifications, measures for dispersal, concealment and camouflage, exploiting natural conditions (such as weather or terrain), and displaying decoys. They also include rapidly repairing damage and recovering air offensive operational capability.[84]

Distinctive Aspects of Chinese Air Offensive Campaigns

A number of aspects of the PLA descriptions of air offensive campaigns are notable. One is that descriptions of how such a campaign should be conducted resemble, to a large degree, the concepts that the USAF employed during the 1980s. This perhaps should not be surprising, however, given that the capabilities China's air forces currently possess, or are in the process of acquiring, resemble, in many ways, those of the USAF in the 1980s. These resemblances include a mixture of third-

---

[83] Zhang Yuliang, 2006, p. 587.

[84] Zhang Yuliang, 2006, pp. 587–588.

generation and fourth-generation multirole and air superiority fighter aircraft, dedicated attack aircraft, and escort and standoff jamming aircraft, but no fifth-generation fighters or stealthy strike aircraft. This being the case, the similarity of employment concepts may simply be the result of the PLAAF, presented with similar challenges and tools as their U.S. counterparts in the 1980s and governed by the same laws of physics, independently arriving at the same conclusions. Alternatively, it is possible that the PLAAF has simply imitated and adopted U.S. doctrine and operating concepts from that period. Most likely, some combination of both processes has occurred.

Despite the distinct similarities to 1980s USAF employment concepts, however, Chinese concepts for conducting air offensive campaigns also exhibit several unique aspects. First is the emphasis on information operations, including its identification in the 2006 edition of *Study of Campaigns*[85] as one of four distinct tasks in an air offensive campaign. As noted earlier, this is consistent with the PLA's concept of informationized warfare since 2004.

Second is the emphasis on surprise, deception, and evasion when conducting air strikes. USAF offensive concepts tend to be based on systematically suppressing and destroying an enemy's air defenses so that subsequent strikes on other targets can be conducted in a relatively benign air defense environment. PLA descriptions of air offensive campaigns, by contrast, do not assume that such a level of air supremacy will ever be achieved and, therefore, that strikes will have to be conducted in a situation in which air defenses are a significant and continuing threat. This view certainly seems realistic given that, unlike the United States in its recent conflicts with Iraq, Serbia, and Afghanistan (or potential future conflicts for which it prepares with Iran or North Korea), China is indeed unlikely to enjoy air supremacy in the primary contingency for which the PLA prepares—a conflict with Taiwan and the United States.

A third distinct aspect is related to this second aspect and is the emphasis on defensive operations even in an otherwise offensive campaign. Just as information operations is identified as one of four dis-

[85] Zhang Yuliang, 2006.

tinct tasks in an air offensive campaign, resisting enemy air counterattacks is another major task of an air offensive campaign. As with the emphasis on surprise, deception, and evasion, however, this is consistent with an assumption that the PLA will not have absolute air supremacy throughout the campaign, but rather will have air superiority only in particular places and times, and therefore will be subject to counterattack. Again, in a conflict with the United States and Taiwan, this assumption is entirely realistic.

A fourth distinctive feature of Chinese concepts for conducting air offensive campaigns is a strong preference for destroying enemy air forces on the ground. This preference is not unique to China—aircraft parked on the ground are virtually helpless and, if attacked there, can be destroyed in large numbers as, for example, Israel did to the Egyptian and Syrian air forces during the Six Day War of 1967. Nonetheless, this preference, and the implied desire to avoid an air-to-air war of attrition, both suggests a recognition that the PLA lacks an advantage against likely adversaries (such as the United States) in the combat capabilities of its fighter aircraft and skill of its pilots and suggests a likelihood that the PLA will seek to achieve strategic or operational surprise against an opponent, so that the opponent's air forces are still on the ground when attacked. The means for attacking aircraft on the ground, moreover, are not limited to aircraft but could include SSMs, SOF, and other types of capabilities.

A final feature of air offensive campaigns worth noting is the apparent assumption that the targets of such a campaign are on land. Coupled with the minimal discussion of air-to-sea combat in PLAAF publications, this suggests that the PLAAF has a minimal role in conducting maritime strike missions. Since maritime strike would be an important mission in a variety contingencies—most especially, an invasion of Taiwan—the likely implication is that maritime strike is primarily a mission of the PLAN aviation forces.

CHAPTER SIX

Air Defense Campaigns

In its thinking on air defense employment concepts, the PLA draws on its own traditions, while adjusting its methods in keeping with what it views as predominant trends in technology and capability. Air defense has historically been the raison d'être of the PLAAF. It has traditionally received the greatest emphasis in doctrinal thought, although, today, belief in the efficacy of offensive air action is growing, and air defense campaigns do not enjoy the pride of place they once held. Nevertheless, air defense operations receive greater emphasis in China than they do in many countries (including those with roughly comparable technology). The PLAAF recognizes that, under some circumstances, it will be necessary to organize its air effort along primarily defensive lines, at least until it can gain the initiative and transition to the offensive.

Just as air offensive campaigns incorporate defensive operations as part of a larger offensive repertoire, air defense campaigns incorporate offensive air operations as well as defensive ones. Nevertheless, given that the centers of gravity in offensive and defensive campaigns are fundamentally different, important aspects of the organization and disposition of airpower are different in the two types of campaigns. The forces involved in offensive campaigns tend to be task organized, whereas defensive campaigns are largely organized geographically. Offensive campaigns call for a relatively forward-leaning deployment pattern, with offensive air units deployed in bases relatively near the enemy, while defensive campaigns call for layered defenses in depth.

Air defense campaigns can, according to Chinese military writings, be national in scope, or they can be confined to a particular theater. Depending on the circumstances, the entire air effort in a given war could be defensive, a single phase could be defensive, or, in the case of a geographically wide-ranging conflict, some theaters could be defensive while others are offensive. In a war over Taiwan, for example, the PLA might conduct an offensive air campaign in the area opposite Taiwan while preparing for air defense campaigns to the north and south in anticipation of possible retaliation or counterattack by U.S. forces. Chinese military writings suggest that, as circumstances permit, commands should seek to transition from defensive to offensive campaigns, though circumstances might also dictate the opposite progression.

General Objectives, Forms, and Methods

Air defense campaigns are conducted within a certain airspace to resist large-scale enemy air attack campaigns. They can be conducted as independent campaigns but are more usually conducted as part of a broader joint campaign. They are designed to "destroy or attrit the enemy's offensive air strength, guarantee the security of important objectives, avoid or reduce the damage from enemy air attack, smash the enemy's offensive plans, and create the conditions for victory on ground, sea, or air."[1] A textbook on military operations lists three primary missions for these campaigns: protecting the capital against air attack, protecting other important targets within the theater, and seizing and keeping air superiority.[2] The relative importance of these missions is changing as the third (seizing and keeping air superiority) is gaining in emphasis relative to the others.

There is apparently little definitive agreement on how air defense campaigns should be categorized. One source suggests that, depending on the scope, these campaigns can be divided into three types: key-area air defense campaigns [要地防空战役], theater air defense campaigns [战区防空战役], and multitheater air defense campaigns

---

[1] See PLAAF, 2005, p. 101.

[2] Bi, 2002, pp. 471–472.

[多战区防空战役].³ Another suggests a somewhat simpler binary division between key-area air defense campaigns and theater air defense campaigns.⁴

Older sources, as well as some recent ones, refer to a wider variety of air defense operations that are not classified as forms of air defense campaigns per se but nevertheless have their own distinct characteristics and requirements. These include, among others, strategic air defense [战略防空], key-point defense [要点防空], key-area defense [要地防空], battlefield air defense [野战防空], mobile ground force air defense [战役军团机动防空], people's air defense [人民防空], national air defense [国土防空], and regional air defense [区域防空].⁵ While these distinctions are sometimes useful (and hence several are still employed selectively by Chinese air force strategists), the general streamlining of categories reflects the maturation of the PLAAF, its rising influence in the defense community, and its determination to impose a better-integrated, simplified, and centralized system of organization in all its activities—including air defense.

While recognizing these distinctions (especially that between key-point and theater air defense) and borrowing some of this terminology for the discussion of special topics, the discussion in this chapter is organized around theater air defense. Although they are important in Chinese employment concepts, key-area and key-point defenses can be regarded as component parts of larger campaigns. Most contemporary Chinese military writings, especially the most official ones, organize their discussions in this way, even when they recognize distinctions

---

³ PLAAF, 2005, pp. 101–102; *Air Force Dictionary* 《空军大辞典》, Shanghai: 上海辞书出版社 [Shanghai Dictionary Press], 1996, pp. 18–19.

⁴ Bi, 2002, pp. 473–474.

⁵ In the chapter on air force strategy found in the 1996 *Air Force Dictionary*, entries are found for air defense, strategic air defense, national air defense, key-area air defense, battlefield air defense, people's air defense, regional air defense, active air defense, passive air defense, and air defense system. See *Air Force Dictionary*, 1996, pp. 7–11. Some newer works draw on these distinctions. In Cui et al., 2002, pp. 143–144, for example, the authors divide combined air defense campaigns into three types: key-point, key-area, and regional defense. They also include a separate chapter on battlefield air defense (pp. 286–307) and one on mobile ground force air defense (pp. 322–340).

between campaign types. And Chinese air force strategists identify a trend in "air defense thought" away from key-area defense and toward large-area defense.[6]

Operations in support of theater air defense are grouped into three types: resistance operations, which include both ground and air maneuvers and fires on attacking enemy forces; counteroffensive operations, which include air and missile attacks on the enemy's bases; and close protection operations, which include cover, concealment, deception, and recovery operations. In keeping with the general trend toward emphasis on offensive action, counteroffensive operations are described as the "decisive form" of air defense. Resistance operations are, however, still called the "basic" or "main" method.[7] Close protection is, meanwhile, said to remain important under the high-tech conditions of 21st-century warfare.[8] In practice, the idea is to combine the early interception of enemy attacks with full-depth, layered resistance and to protect targets and forces while gradually increasing the tempo of counterattacks on enemy bases.[9]

Characteristics and the "Crisis in Air Defense"

The PLAAF's formula for air defense campaign characteristics reflects an assessment—one that has strengthened in the past decade—that air defense campaigns are extraordinarily difficult to wage successfully and that they place the defender in a reactive position, surrendering the initiative to the enemy. The Chinese discussion of these characteristics reinforces other indicators that the PLAAF has significantly

---

[6] See Wang Fengshan [王凤山], Yang Jianjun [杨建军], and Chen Jiesheng [陈杰生], eds., 《信息时代的国家防空》 [*National Air Defense in the Information Age*], Beijing: 航空工业出版社 [Aviation Industry Press], 2004, p. 119. See also "Trends in Air Defense Thought" later in this chapter.

[7] Cui et al., 2002, p. 142.

[8] Cui et al., 2002, pp. 215–217.

[9] This formulation is taken from Lu, 2004, pp. 267–268. Similar language is found in other sources.

downgraded the status of defensive campaigns in its doctrinal thinking. Summarizing this situation, one group of PLAAF strategists (and proponents of the defense) has referred to a "crisis in air defense."[10]

An authoritative Chinese reference book on military operations lists five characteristics associated with air defense campaigns. First, "the lead time to prepare for war is short, and it is easy to fall into a reactive position" [临战准备短促，易陷入被动].[11] This is, the book says, a critical characteristic that separates air defense campaigns from all others, and is especially true at the start of the campaign. Because the attacker holds the initiative in launching the campaign, it can meticulously prepare its activities and can select the means of attack, the methods and weapons, and the most advantageous time and location to strike. Because warning and preparation time are short, the defender must prepare for the campaign while entering combat.[12]

The second characteristic is a "large defense area, and heavy and numerous responsibilities" [防卫空间广阔，任务繁重].[13] With the range of aircraft having increased dramatically, the reference book says, many now have transcontinental range. The battlefield, which can now encompass several theaters and several million square kilometers, has grown proportionately. Within this space, air defense commanders must not only organize active resistance and covering operations and defend important military, political, and economic targets; they must also organize large-scale counterattacks against the enemy's air bases and missile-launch platforms.[14]

The third characteristic is the "diversity of participating forces and complexity of coordination" [参战力量多元，协同复杂].[15] The *Study of Campaigns* documents this, listing the diverse forces that must be

---

[10] There is an extended discussion of the "crisis in air defense," which is said to have begun during the 1980s, in Wang Fengshan, Yang Jianjun, and Chen Jiesheng, 2004, pp. 29–35.

[11] Zhang Yuliang, 2006, pp. 602–605.

[12] Zhang Yuliang, 2006, pp. 602–603.

[13] Zhang Yuliang, 2006, p. 603.

[14] Zhang Yuliang, 2006, p. 603.

[15] Zhang Yuliang, 2006, pp. 603–604.

coordinated: PLAAF fighters, SAM, AAA, EW, and appropriate support forces; naval fighter aviation; army and navy SAM, AAA, and EW units; and people's air defense AAA elements. In the course of operations, coordination would be required between the PLAAF and the other services; within the PLAAF between its component parts (e.g., ground-based and fighter elements), as well as between units of the same or different types; between the services and people's air defense elements; between hard-kill and soft-kill (e.g., EW) elements; between operational elements and support units; between the various theaters and air defense districts; and between resistance operations and counterattack ones. All of this complexity, the source suggests, "helps drive the reactive nature of [air defense] operations."[16]

The fourth characteristic is that "the information fight will be intense and will last from start to finish" [信息领域争夺激烈，信息对抗贯穿战役始终].[17] In a general sense, according to this book, the information struggle has already become an important operational action in air attacks and defense. Presumably in reference to the U.S. military, this source also suggests that "the enemy will exploit his technological superiority to first employ information weapons against our intelligence and warning systems, command centers, and communications, and follow-up with all kinds of hard- and soft-kill attacks to suppress our air defense system." The imperative on the defense, then, is to deploy reconnaissance and early warning systems early and identify threats as they develop.[18]

Finally, the fifth characteristic is that there will be a "fierce struggle between systems, and mixed offensive and defensive operations" [系统对抗激烈,攻防交织进行].[19] In addition to organizing fighters, SAM, and AAA elements to conduct continuous attacks against inbound aircraft, "it is necessary to concentrate elite, long-range forces to counterattack enemy airfields and sea platforms." In the future, the

---

[16] Zhang Yuliang, 2006, p. 604.

[17] Zhang Yuliang, 2006, p. 604.

[18] Zhang Yuliang, 2006, p. 604.

[19] Zhang Yuliang, 2006, p. 604.

reference continues, "there will be offensive operations in the defense, defensive operations in the offense, a mixing of offense and defense, and, overall, a fierce struggle."[20]

The discussion of air defense campaigns in Zhang Yuliang, 2006, has a significantly more pessimistic tone about the prospects for defensive success than the assessment found in a 1996 reference work (*Air Force Dictionary*). The *Air Force Dictionary*, like Zhang Yuliang, 2006, cites "short preparation time" as a defining characteristic but does not follow up with the argument that this necessarily places the defender in a reactive position.[21] Rather, the *Air Force Dictionary* states that, "Since the 1960s . . . new offensive and defensive weapons have been employed, and the struggle between air attack and air defense has become more fierce."[22] This may reflect a battle between ever more capable fighter aircraft and SAM missiles, but it hardly suggests the demise of the defense.

The differences between the 2000 edition of *Study of Campaigns* and the more recent version are subtler in their assessment of the offense/defense balance, but nevertheless noticeable.[23] In Zhang Yuliang, 2006, for example, the term *reactive* [被动] (sometimes translated as "passive") appears more prominently and more frequently in reference to air defense campaigns than it does in Wang Houqing and Zhang Xingye, 2000.[24] Whereas the discussion of each component characteristic in Wang Houqing and Zhang Xingye, 2000, ends with a suggestion about how to overcome the obstacles or challenges listed, the Zhang Yuliang, 2006, discussions simply conclude with a summary statement of the difficulties. A 2004 study edited by a PLAAF strategist with a background in SAM units asks whether recent Chi-

---

[20] Zhang Yuliang, 2006, pp. 604–605.

[21] *Air Force Dictionary*, 1996, pp. 18–19.

[22] *Air Force Dictionary*, 1996, p. 19.

[23] Wang Houqing and Zhang Xingye, 2000; Zhang Yuliang, 2006.

[24] For example, in Zhang Yuliang, 2006, "reactive" appears in the descriptor for the first characteristic, whereas it appears only in the body of the text in Wang Houqing and Zhang Xingye, 2000. In Zhang Yuliang, 2006, it also appears as an implication of "command complexity," while it does not in Wang Houqing and Zhang Xingye, 2000.

nese commentary on a "crisis in air defense" is accurate. Although the answer provided suggests that the "crisis" may be overcome, even this relatively optimistic source acknowledges a wide range of challenges.[25]

Trends in Air Defense Thought

How do the Chinese intend to address the challenges in conducting air defense outlined in this chapter? Before going on to discuss specific provisions found in Chinese military publications on disposition, command, and activities associated with these campaigns, it is worth reviewing a more general discussion of China's "air defense thought." One such discussion is particularly rich in detail and clearly informed by changes in Chinese doctrine and force structure.[26] Although that discussion overlaps with the propositions about the characteristics of air defense put forward in the 2006 text described earlier, it is not identical and is more solution oriented. It is, therefore, worth considering in some detail. This source identifies six broad trends in air defense thought.[27]

First, the importance of "key-point defense" [要点防空] is waning, while that of "large-area defense" [大区域防空] is growing.[28] Given the PLAAF's historical focus on the defense of cities and other key sites, this represents perhaps the biggest change in thinking. Because of increased standoff ranges, the thinking goes, attackers must be engaged earlier and farther from the target. The only way to do this is to expand the battle space. The forward edge of the aerial battle space must be pushed toward the enemy, and interception must occur earlier, even as the bulk of resistance operations will be conducted in depth. Given the increased range of aircraft and their ability to refuel in flight,

---

[25] Wang Fengshan, Yang Jianjun, and Chen Jiesheng, 2004, pp. 29–35.

[26] Wang Fengshan, Yang Jianjun, and Chen Jiesheng, 2004, pp. 113–122.

[27] Taken from Wang Fengshan, Yang Jianjun, and Chen Jiesheng, 2004, pp. 113–122. We have reordered the trends.

[28] Wang Fengshan, Yang Jianjun, and Chen Jiesheng, 2004, pp. 119–121.

"forward" defense must be multidirectional and cover all approaches, rather than unidirectional or linear.

Second, there is a move away from fixed defenses toward "mobile air defense" [机动防空].[29] The ability to "shoot and scoot" can improve survivability in the face of more-effective offensive reconnaissance and attack capabilities. Mobility can help plug holes or weak links in the air defense network and create conditions for destroying the enemy by creating local superiorities. Mobility is receiving new emphasis in Chinese military publications, but Chinese military writings emphasize that mobility in air defense is not new to the PLAAF. It is, rather, regarded as a historical strength, with the defensive antiaircraft ambush said to have been pioneered by PLA SAM forces.[30]

Third, exclusively defensive air defense is giving way to a concept of "offensive air defense" [攻势防空], driven by the increasing effectiveness of offensive operations.[31] Given that resistance operations are still the "basic form" of air defense, the more salient point may be the increased prominence of "integrated attack and defense" [攻防结合], in which the offense is used to assist the defense. In part, the inclusion of offensive action is meant to capitalize on the purported effectiveness of airfield attack, and, in part, it is meant to gradually seize the initiative from the enemy. Commanders should, one source suggests, "actively organize counterattack operations of various scales to distract and attrit the enemy, disrupt his plans, destroy his offensive posture, gradually move the enemy into a reactive mode, and, ultimately, seize the operational initiative."[32]

The fourth trend is toward "information air defense" [信息防空].[33] As noted previously, information has become a core component of battle strength, and "gaining information superiority must be incorporated during the entire course of the air defense campaign."

---

[29] Wang Fengshan, Yang Jianjun, and Chen Jiesheng, 2004, pp. 117–119.

[30] On antiaircraft ambushes, see PLAAF, 2005, p. 134.

[31] Wang Fengshan, Yang Jianjun, and Chen Jiesheng, 2004, pp. 115–116.

[32] Wang Fengshan, Yang Jianjun, and Chen Jiesheng, 2004, p. 116.

[33] Wang Fengshan, Yang Jianjun, and Chen Jiesheng, 2004, pp. 113–115.

The fifth trend is the unification of air and space defense [防空防天一体化].[34] This largely translates into the need for integrated command and control and the understanding that, "at the start of the 21st century, whoever controls space, controls the planet."[35] Space is, in short, seen as the new high ground.

Sixth is the trend away from single-service air defense and toward joint air defense—part of the larger trend in PLA thinking toward the use of joint operations.[36]

These trends are apparent in many shifts in Chinese military writings and in the organization and deployment of the force. There is, however, some debate about the relative importance of various trends and how they should be reflected in doctrine and organization. Even where there is general agreement, moreover, military organizations cannot turn on a dime, and there are many elements of old mixed with new. Despite the new emphasis on large-area defense in Chinese military writings, much of the force remains organized to defend key points, particularly cities. A large percentage of air defense assets, for example, including three composite (SAM and AAA) air defense divisions, are tied down defending Beijing.

Similarly, despite doctrinal emphasis on large-area defense, the zones into which Chinese air defense commands are broken follow political (provincial and urban) boundaries, rather than militarily relevant geography. And despite the new emphasis on "informationized air defense," the primary principle for command and control is procedural, rather than active control. In other words, responsibilities are divided geographically and temporally (into phases), rather than employing air defense forces in an integrated manner according to incoming data on threats and opportunities.

These trends in PLA thinking on air defense, therefore, tell us more about the organization's aspirations and the direction it may be headed than about where it is today. In many cases, the new thinking may take one or more decades to move from concept into practice.

---

[34] Wang Fengshan, Yang Jianjun, and Chen Jiesheng, 2004, pp. 121–122.

[35] Wang Fengshan, Yang Jianjun, and Chen Jiesheng, 2004, p. 121.

[36] Wang Fengshan, Yang Jianjun, and Chen Jiesheng, 2004, pp. 116–117.

Nevertheless, it has already begun to have an impact—albeit uneven—on organizational patterns and training.

Command Arrangements and Coordination

In an air defense campaign, there are three levels of command. At the top rests the theater air defense command [战区防空指挥机构]. Given the new Chinese emphasis on flexibility and joint action, theater air commands may encompass more than one peacetime MR, but, in these cases, we would anticipate that a single commander, possibly one of the MR commanders, would be given the lead and authorized to establish the larger theater air defense command. The theater air defense command is charged primarily with defining the subordinate air defense zones, commanding trans-zone operations, and establishing the distribution and (if need be) the redistribution of assets between zones.

Below the theater level is the "air defense zone" command [防空分区指挥机构]. Air defense zone commanders coordinate the relevant units and assets of different services within their zone, including PLAAF aviation units, SAM and AAA units belonging both to the PLAAF and PLA Army, naval aviation and AAA units, SAMs on naval ships, and the relevant militia and reserve units assigned to particular areas. The air defense zone command will assume overall command and coordination (e.g., methods, timing, routes) of air defense operations within the zone beyond those restricted strictly to key- (or strategic) point defense. Although it is possible that boundaries could differ during wartime, zone boundaries during peacetime appear to correspond to political, civilian administrative boundaries.[37]

Below the zone level are "key-area" air defense commands [要地防空指挥机构]—primarily urban or industrial areas, though they may include other key locations (e.g., air and naval bases). Defense of key areas can be assigned to the ground, naval, or PLAAF units

---

[37] See, for example, the hierarchy of sub-MR air defense zones found on the Wuxi (undated) air defense website.

that occupy them. In the case of urban defense, militia, reserve, and civil defense elements also provide important assets. Despite statements that imply the flexible use of multiservice assets within key-area air defense commands, a variety of more-specific statements suggest that army and navy air defense capabilities will be effectively controlled by their parent organizations (though they will be expected to coordinate with the zone air defense headquarters in which they reside). The expectation is apparently that PLAAF capabilities may be dispatched to supplement key-area defenses where inadequate, but not the other way around.[38]

As the discussion of trends suggests, air defense campaigns are usually joint operations involving not just the PLAAF but also the PLA Army or PLAN, though they may, in some very limited cases, also be fought exclusively by the PLAAF. As in the other campaigns, coordination and control are heavily procedural. When air and ground elements are both assigned to these campaigns, coordination "is usually dominated by air operations" and accomplished by means of zone, direction, and altitude. In the defense of key areas or points [要地 or 要点], defending fighter aircraft often patrol at the greatest distance in front of the area being defended, while SAM and AAA units defend closer to it, with responsibilities divided by altitude.[39]

Disposition of Forces

In a theater-level air defense campaign, forces are disposed in three lines: a first-line interception zone [一线拦截区], a second-line blocking and destruction zone [二线阻歼区], and a deep covering zone [纵深掩护区].[40]

The first-line interception zone is situated closest to the enemy and pushed as far forward as possible. A relatively small number of

---

[38] Cui et al., 2002, pp. 314–315.

[39] Bi, 2002, p. 479.

[40] Except where otherwise noted, the information on disposition of forces is from Bi, 2002, p. 478.

fighters and long-range ground-based air defense units are situated in this zone. The primary mission for units located here is to find and engage enemy attackers as far out as possible, destroying them if possible, or disrupting their formation, and passing the targets on to units located in the second zone if not. Chinese writings point to the need for an integrated, overlapping radar network to manage early warning and emphasize pushing the battle space outward by stationing air patrols forward (beyond the range of ground-based radar if possible and necessary) and by deploying long-range SAM and high-altitude AAA as far forward as possible. The use of offshore islands and ships (including civilian ships) for radar is also encouraged.[41]

The second-line blocking and destruction zone is located in the middle area of the theater and is the "operating zone for comprehensive firepower." Its forward edge is defined by the areas where full operational support can be well organized and where the integrated and overlapping use of short-, medium-, and long-range systems becomes possible. The bulk of fighter units and SAM and AAA units are deployed in this zone. Their mission is to conduct a layered defense against enemy targets and to launch continuous attacks designed to destroy or block their operations.

Finally, the deep covering zone covers the remainder of the battle space. It is organized around the terminal defense of key points, including civilian and military targets, and includes small numbers of fighters and SAM and AAA systems. Reserve forces, counterattack forces, and support aircraft (including EW, early warning, and refueling aircraft) are located here.

Several differences between the deployment for air defense campaigns and air offensive campaigns are apparent. First, air defense deployments are less front-loaded. Only small numbers of aircraft are deployed in the front line. Second, although offensive action is part of the defensive campaign, the aircraft involved in such attacks (bombers, attack aircraft, and fighter-bombers) are located in the deep covering zone, well away from the action, and must deploy forward to execute their missions. Third, the defensive disposition facilitates con-

---

[41] Cui et al., 2002, p. 313.

tinuous attacks on incoming aircraft, rather than concentrated strikes on enemy aircraft. In large measure, the difference derives from a desire to disrupt the attacker and protect key facilities, rather than destroy the attacker outright (though obviously that objective will be pursued when possible).

Air Defense Campaign Operations

Operations conducted as part of an air defense campaign are organized into the categories of resistance, counterattack, and close protection.

Resistance Operations

Resistance operations are regarded as the "basic" or "main" type of defensive operation and include six categories of operations: distant intercept and attack [尽远截击], echeloned resistance [梯次抗击], maneuver ambush [机动伏击], hunting and destruction attacks [游猎歼击], obstacle and blocking operations [设障阻击], and destruction of the enemy's attack structure [结构破坏].[42] The six types of operations are not necessarily mutually exclusive. Maneuver ambushes and hunting and destruction attacks can be part of an echeloned attack, and destruction of the enemy's flight structure is an objective sought in all forms of resistance operations. Nevertheless, the list provides a structure for discussing the mechanics of resistance activities in air defense campaigns.

---

[42] Cui et al., 2002, pp. 210–213. Note that this source provides the most-expansive and best-organized discussion of air defense operations, but the list is not an official one. It is broadly consistent with other sources but not replicated in them. PLAAF, 2005, p. 101, for example, emphasizes "consistent preparation," "accurate detection," "long-distance and persistent engagement," "counter-attack on enemy airfields," and "protection of equipment," the last three of which appear to correspond to the "resistance," "counterattack," and "close protection" operations specified in Cui et al., 2002. It does not, however, describe these principles as explicit elements of an air defense campaign, but rather mentions them as the primary recommendations for the handling of an air defense campaign. Similarly, Bi, 2002, pp. 267–268, provides recommendations that include "detection," "force adjustment," "proper command organization," and exploitation of detection systems all as aspects of resistance in an air defense campaign.

Distant Intercept and Attack. Distant intercept and attack is used to expand the battle space and is employed against all types of attackers, but especially enemy aircraft with standoff attack capabilities (e.g., bombers with cruise missiles). The objective is to destroy the enemy far from critical objectives, or, if that proves impossible, to disrupt enemy cohesion and pass the targets on to elements in the second line of defense. Distant intercept is conducted by small, high-quality fighter units and long-range SAM units deployed as far forward as possible. Fighter units may be deployed to first-line "battlefield" air bases [一线野战机场], reserve airfields, and highways converted to airstrips, using dispersion and frequent movement to improve survivability. SAMs and high-altitude AAA may be deployed on offshore islands. They may also be deployed on civilian ships that cruise in coastal waters seeking opportunities to ambush enemy aircraft.[43]

Echeloned Resistance.[44] This involves the use of fighter aircraft and all kinds of ground-based air defense weapons to launch continuous [连续] attacks on enemy forces. In organizing ground-based weapons, systems will be "scientifically paired" to create mutually supporting fires and a network of integrated high-, medium-, and low-altitude fires. (Notable here is the lack of mention of "scientific" or any other pairing between fighter and ground-based defenses, which operate in different areas.) Defending forces will employ different attack methods, depending on circumstances. When faced with electronic interference or poor visibility, for example, AAA weapons may employ "barrage" [拦阻射击] attacks, in which a likely avenue of approach is saturated at all altitudes with fire.

Maneuver Ambush.[45] Maneuver ambushes are sudden and unexpected fire attacks on enemy forces by mobile SAMs and high-altitude AAA systems. This type of attack is consistent with the PLAAF's view

---

[43] Cui et al., 2002, p. 211.

[44] This is discussed in Cui et al., 2002, p. 211, and PLAAF, 2005, p. 144.

[45] The discussion of maneuver ambushes is drawn equally from Cui et al., 2002, pp. 211–212, and PLAAF, 2005, p. 135. *Maneuver ambush* has an entry in PLAAF, 2005. In PLAAF, 2005, the Chinese term employed is 机动设伏, rather than 机动伏击, the term employed in Cui et al.

of the importance of mobility in air defense operations, but also derives from traditional PLA air defense strengths. Indeed, Chinese military publications claim that the PLA pioneered maneuver air defense attacks and that they have historically proven highly effective. "Between 1962 and 1972," says the *China Air Force Encyclopedia*, "SAM forces established more than 100 maneuver ambushes, traveled some 200,000 kilometers taking them across 20 provinces, independent cities, and autonomous regions, and shot down eight aircraft."[46] In 1979, SAM units, employing maneuver ambush "in the southwest" (presumably against Vietnamese targets) damaged and shot down aircraft that had crossed the border.[47]

There are two types of maneuver ambushes: the waiting ambush [待伏] and the induced ambush [诱伏]. In the case of the induced ambush, deception will be employed, such as the use of fake targets for the enemy on reserve airfields or fake radar transmissions to lure enemy units to the area. Key targets for the maneuver ambush are support aircraft, like AWACs, tankers, and EW aircraft.

Hunting and Destruction Attacks.[48] Hunting and destruction attacks are conducted by elite, specially trained fighter elements [分队] or individual aircraft. Unusual in the context of the PLAAF's tendency toward tight control and scripting, elements assigned to this task are given a general intent and an area of operation [作战意图], but the element commander is given freedom to conduct his or her own planning or make changes to operations under way. Typically, the element will attempt to remain undetected while positioning itself along the flanks of expected ingress or egress routes or near enemy air bases. Targets may include aircraft that are taking off or landing, refueling, in transit to or from bases; aircraft that have been damaged and left

---

[46] PLAAF, 2005, p. 135.

[47] Yunnan, bordering Vietnam, is considered part of China's southwest.

[48] The discussion of hunting and destruction attacks [游猎歼击] is based primarily on Cui et al., 2002, p. 212, and PLAAF, 2005, p. 131. In PLAAF, 2005, the Chinese phrase employed is 空中游猎 and is translated by the encyclopedia as "air hunting." The encyclopedia provides an additional entry for "air sweep" [空中游击], for which it simply references the entry for "air hunting."

their formation; or high-value support aircraft (e.g., AWACS, tankers, EW aircraft).

Establishing Obstacles and Blocking. Blocking enemy attacks is accomplished with a combination of physical obstacles, including balloons with steel cables and suspended mines, and saturation fire attacks by AAA to destroy incoming aircraft or cruise missiles at low altitude. Cruise missiles are regarded as a major threat to the integrity of air defense and national command but are also seen as having vulnerabilities (including slow speed and set patterns of attack) that can be exploited by clever defenders.[49]

Destroying the Enemy's Flight Structure. Although listed as a separate type of operation in Cui et al. (2002), the destruction of the enemy's formation or integrity is really more of an objective or priority that is incorporated into all other types of resistance operations. The emphasis here is on the destruction of critical force multipliers, like AWACS, EW aircraft, or tankers, though it can also include the disruption of larger attacking formations through continuous attack or other methods.

Counterattack Operations

In keeping with the concepts of "offensive air defense" and "integrated attack and defense," counterattack operations will be launched against enemy air bases—including aircraft carriers—during air defense campaigns. These may include aerial surprise attacks [空中奇袭], firepower attacks [火力突击], rear-area raids [敌后破袭], and maritime surprise attack [海上偷袭].[50]

Aerial surprise attacks are similar in structure and organization to attacks carried out during air offensive campaigns, but also have distinctive features. Due to the fact that, in defensive campaigns (unlike offensive ones), counterattack forces are deployed in the deep covering zone, counterattacks require initial staging forward. Also, given the assumption that the enemy will hold the overall advantage in equipment and initiative (hence necessitating a defensive campaign), the

---

[49] Cui et al., 2002, pp. 212–213.

[50] Cui et al., 2002, pp. 213–215. On counterattack operations, see also Bi, 2002, p. 481.

use of stealth and deception is particularly important. Attacks will make use of night, complex weather, terrain masking, and low-altitude approaches to achieve their objectives.[51] Attacks may also be executed by small units employing sequential sorties, rather than by the more-concentrated formations that might be preferable under other circumstances. At least one source suggests that China's air forces might attempt to infiltrate enemy formations on their return flights to penetrate the enemy's air defenses.[52]

Firepower attacks on enemy air bases, support facilities, and aircraft carriers would be executed by tactical missile units of the PLA Army, the PLAN's coastal missile units, or the conventional missiles of the Second Artillery. Again, given the assumption of enemy air superiority inherent in an air defense campaign, the need for stealth and, especially, repositioning after the strike receives particular emphasis. The scale of attacks may be smaller than those contemplated in the context of offensive air campaigns. And the objectives will be more likely to include disruption than destruction, in order to gradually wrest the initiative away from the enemy.[53]

Rear-area raids on airfields, support facilities, or command headquarters may be conducted by airborne forces or SOF. These raids may take the form of firepower attacks using mortars or other light artillery, or they may take the form of infiltration designed to place explosives or destroy equipment. Although airfield raids or seizures are part of the U.S. airborne and special forces repertoires as well, the likelihood that these will be incorporated into the overall air plan is particularly high in the Chinese case, given the inclusion of these elements (both airborne forces and SOF) in the PLAAF, as opposed to them belonging to another service.

Maritime surprise attacks will employ submarines, missile boats, tactical missiles, coastal missiles, and aviation units to attack aircraft carriers, targets of particular concern to Chinese air defense planners. Since the PLAAF appears to have limited maritime strike capabilities,

---

[51] Cui et al., 2002, p. 213.

[52] Cui et al., 2002, pp. 213–214.

[53] Cui et al., 2002, p. 214.

however, these attacks would probably be carried out largely by PLAN assets.

Close Protection Operations

Close protection operations are designed to prevent or limit damage to cities and facilities and to ensure speedy recovery from damage that does occur. They take the form of fortification protection [工事防护], dispersal and concealment [流散隐蔽], careful camouflage [严密伪装], and extensive mobility [广泛机动]. These are traditional PLA strengths (at least in the first three cases) that have gained in importance with the advent of more-effective aerial attack systems.[54]

Fortification protection involves the use of engineering to build shelters, preferably underground. PLA planners have studied the efforts of Serbian, Iraqi, and other militaries to protect critical facilities. They acknowledge that air strikes have become more accurate and potent. However, they also believe that the defensive measures employed by these countries (1) were relatively successful from a tactical perspective (in, for example, protecting some aircraft from destruction), and (2) could be further improved.[55] Dispersal and concealment are intended to reduce the vulnerability of personnel and equipment and are deemed important in the defense of urban centers. Camouflage combines traditional means of concealment (e.g., camouflage paint and smoke) with newer elements (e.g., anti–infrared, anti–laser ranging).[56]

Extensive mobility is the newest aspect of close protection and is predicated on the understanding that precision weapons make the defense of fixed sites difficult. Consequently, PLA publications emphasize that aviation units should take full advantage of reserve and civilian air bases and that SAM and AAA forces should reposition to prepared sites often. PLA commentators appear to believe that, particularly in the case of SAM elements, the fortification of secondary and tertiary field positions is currently inadequate and that greater engineering

---

[54] Cui et al., 2002, p. 215. On close protection operations, see also Bi, 2002, p. 481.

[55] Peng and Bi, 2001, pp. 156–178, 184–218; Cui et al., 2002, p. 215.

[56] Cui et al., 2002, pp. 215–217.

efforts should be made to allow SAM systems (especially older, less mobile ones) to "shoot and scoot."

Defense of Cities and Bases

Defense of key areas [要地] (which include primarily urban areas) and key points [要点] (which include more-specific targets, such as industrial centers or military bases) is also part of air defense campaigns.[57] These bastions will be located throughout the depth of Chinese airspace and contribute to the larger theater air defense (there are, in this aspect, however, points of tension with trends in Chinese thinking on air defense).

The forces involved in the defense of urban areas include an intelligence and early warning group, an electronic countermeasure group (including aircraft and local ground units), an air defense group (including fighters, SAMs, and high-altitude AAA), a counterattack force (including attack aircraft and bombers), a ground defense force (primarily engineers and transportation repair workers), and a support force (including command and logistics elements). These elements will, of course, be created as needed, and it is doubtful that a counterattack force would be assigned to the defense of a single urban target except under particular (and unusual) circumstances.[58]

The employment of these forces calls for expanding the battle space by deploying fighter aircraft forward in dispersed and hidden bases and by maintaining, where possible, airborne patrols. SAMs and AAA will be deployed farther back, both in the suburbs and in the city itself. They will generally deploy in a fan shape in front of the city, with emphasis on particularly likely avenues of approach (e.g., along routes where enemy aircraft might expect to make concealed approaches). AAA may maximize fields of fire by positioning themselves on rooftops. As in larger air defense operations, emphasis will be placed on dis-

---

[57] Except where otherwise noted, this section derives from Cui et al., 2002, pp. 266–285.

[58] Cui et al., 2002, pp. 274–275.

persion and mobility—including, most importantly, shoot-and-scoot tactics designed to enhance survival.[59]

To a significant extent, thinking on these topics reflects a legacy of past practice and tradition. The defense of fixed points may tie down forces and impede the flexible use of assets. In principle, the theater or air defense zone commanders are allowed to allocate assets as needed. But there appears to be a presumption that urban targets will be protected and that a full array of assets (including aviation elements) will be assigned to the task. Fully three integrated air defense divisions, for example, are assigned to the defense of Beijing. To the extent that enemy aircraft or cruise missiles pass near a variety of these bastions, their defense contributes to the larger goal of "echeloned defense" and "continuous attack," but China's urban geography and military basing structure are not necessarily arranged in keeping with the logic of air defense.

Ground (and Naval) Forces, Maneuver Corridors, and Local Superiority

Chinese military publications on air defense divide the discussion of protecting ground forces into two parts: general air defense of battle-field forces [野战防空] and air defense for ground forces' mobile operations [战役军团机动防空]. In the first case, air defense forces may include army, air force, and naval assets, as well as local and militia forces. Army air defense forces will represent the "basic" or "core" force. Air force and naval elements will provide "rapid maneuver" capabilities. And local and militia forces will represent "supporting strength."[60] While higher headquarters may assign air force assets down to support field forces, PLA Army air defense will be relatively self-sufficient when conducting static operations, and, although it will have organic air

---

[59] Cui et al., 2002, pp. 276–277.

[60] Cui et al., 2002, p. 298.

defense assets, it will also rely heavily on passive defenses (e.g., deception, camouflage).[61]

Two methods of protecting ground force mobility are the use of air defense corridors [防空走廊] and local air superiority [局部制空权].[62] In both cases, theater commanders take responsibility for establishing the location, timing, and forces to be used. In part, the need for these methods derives from the increasing vulnerability of moving ground forces to air attack. PLA thinkers, though, also observe that aerial interdiction has historically been a priority of U.S. airpower—and a particularly debilitating one to the United States' enemies. Finally, the PLA has employed these two measures (corridors and local air superiority) with some success historically.

Air defense corridors are established along planned ground force maneuver routes (or supply lines).[63] These corridors are defended by SAM and AAA units. Within the corridor, commanders establish strongpoints at potentially vulnerable chokepoints: transportation hubs, bridges, or narrow gaps in terrain. The defense of these strongpoints rests with the same combination of comprehensive ground-based firepower used in more-generic key-point defenses (i.e., low-, medium-, and high-altitude and short-, medium-, and long-range systems). Defenses at strongpoints are generally arrayed in a fan-shaped disposition when the direction of attack can be anticipated and a circular one when it cannot. A mix of fixed and mobile defenses will be employed, with the most-mobile systems assigned to reinforce weak areas, respond to changes in enemy targeting, and concentrate near friendly ground forces on the move.

Efforts to gain local air superiority in the face of overall superior enemy forces may also be employed to facilitate ground maneuver.[64] The Soviet and Chinese efforts in "MiG Alley" (between the Chongchon

---

[61] Cui et al., 2002, pp. 324–329.

[62] These are discussed in Cui et al., 2002, pp. 329–335. Although these are not mentioned as separate types of operations in PLAAF, 2005; Bi, 2002; or Zhang Yuliang, 2006, both are consistent with general principles found in those sources.

[63] Air defense corridors are discussed in Cui et al., 2002, pp. 329–333.

[64] Seizing local air superiority is discussed in Cui et al., 2002, pp. 333–335.

and Yalu Rivers) during the Korean War provide historical reference for these operations. Efforts to achieve local air superiority combine the use of fighter aircraft and ground-based defenses in "seamless" cooperation (which would probably be accomplished by spatially dividing the battle area). PLA commentators suggest that modern technological conditions make maintaining local superiority more difficult (though also more important). Hence, the scope and timing must be chosen carefully in accordance with the timing and nature of the ground operation being protected, the combat radius and capabilities of friendly aircraft, enemy capabilities, and the ability to sustain operations.

Summing Up the Parts: What Would an Air Defense Campaign Look Like?

Gaining a clear overall mental image of what an air defense campaign might look like requires comparing employment concepts for its various component parts with an assessment of the available forces to execute the campaign, and a sense of the specific political and military parameters in which the campaign would occur. Without specifying the third of these (i.e., the specific conflict scenario), a few general observations about the broad outlines of air defense campaigns can nevertheless be ventured.

First, the bulk of ground-based air defense assets and some aviation assets will be organized for the defense of key points and areas—primarily cities, industrial zones, military bases, and command centers. The forces devoted to these defenses will include the preponderance of AAA systems and less-mobile SAM systems, along with more-modern SAM systems and aviation assets in the case of high-value targets. Although PLAAF employment concepts now emphasize mobile operations, forces for the defense of key points and areas represent the basic building blocks around which other forces would be operated and will likely absorb the largest share (quantitatively speaking) of relevant assets.

Second, there will be a relatively small-scale forward battle, into which PLAAF commanders will continuously feed new assets to replace losses. The forces dedicated to this battle will include relatively advanced, mobile, long-range ground-based systems, as well as high-quality aviation assets. Their mission will be to find and fix (and, if possible, destroy) incoming attackers and, more importantly, win the information war. The forces dedicated to the forward battle will be crucial to acquiring information on attacking forces.

Third, campaign commanders will employ their most-mobile and most-capable assets for counterattacks, for mobile resistance operations (e.g., hunting and mobile ambush missions), and for the establishment of maneuver corridors or the seizure of local air superiority. Although the number of advanced aircraft and SAMs capable of taking part in these types of operations is increasing, demand will likely dramatically exceed supply—a problem that will be further exacerbated by pressures to also devote a portion of these assets to supplement the defense of key points and ground units.[65] Nevertheless, available forces can be supplemented by second-line elements, and PLAAF commanders will likely fight to marshal forces for a counterattack campaign—though one that is smaller in scale than they might prefer.

Distinctive Aspects of Chinese Air Defense Campaigns

A number of aspects of Chinese air defense employment concepts appear distinctive, or at least set them apart from U.S. concepts.

Before discussing these distinctive aspects, however, a number of caveats should be stated. First, as noted in Chapter One, Chinese military publications examined for this study may not reflect actual current practice of the PLAAF. Second, differences may be more of degree and emphasis than of kind. The difference is in how much emphasis each of these receives in force employment guidelines. Third, some (if not much) of the difference may be explained by the material realities facing Chinese and U.S. military planners. (For example, if the United

---

[65] In a confrontation, China's apparent commitment to defend key points, particularly Beijing, could be exploited by Beijing's enemies through launching token attacks against urban targets and thereby tying down air defense resources that might otherwise be redeployed elsewhere.

States were facing a technologically and materially superior foe, as China is, it might also place greater emphasis on air defense.) Finally, Chinese concepts of employment for its air forces are in transition, and capturing their present state can be difficult.

Nevertheless, with these caveats in mind, there appear to be several distinctive elements of China's air defense employment concepts. The first has less to do with its approach to air defense than to the relatively high importance accorded it. This may appear an anomalous observation given the already observed shift toward more–offensively oriented operational concepts. Chinese military publications, however, still give greater recognition to the importance of the defensive than do those of the U.S. military. In part, the PLA's continuing material and technological weaknesses explain the difference, as it might well be forced onto the defensive in the event of war with the United States. Nevertheless, it is notable that the weight given air defense operations is greater than that found in the air forces of other developing states. The legacy of "People's War," combined with the historical subordination of the PLAAF to the PLA Army, is another factor that may help explain this feature of Chinese employment concepts.

Both material factors and historically rooted organizational culture also play a significant role in shaping other distinctive aspects of Chinese air defense employment concepts. As in other types of air campaigns, coordination is primarily based on preplanned procedures rather than being conducted dynamically. In the case of air defense campaigns, procedural control translates primarily into geographic division of responsibilities, with aircraft and ground-based defense forces generally assigned separate sectors. (In counterattacks and other special cases, coordination may be temporal, following timetables or sequences.) Material limitations, particularly the weaknesses in identification, friend or foe (IFF) systems and data links, are probably the primary drivers here.

The balance between material and historical/organizational drivers may be different in other aspects of Chinese air defense employment concepts. The Chinese emphasis on achieving local air superiority and on mobile ambush tactics is clearly shaped by China's likely inability, under many circumstances, to achieve general air superiority.

Yet, these specific responses to the dilemma are also clearly shaped by China's historical experience. Air ambushes were, for decades, a staple of China's response to violations of its airspace by a variety of powers—including U.S. aircraft during the Vietnam War.[66] Efforts to achieve local air superiority were practiced during the Korean War in the so-called "MiG Alley."[67] Clearly, some of these tactics have received little rehearsal during China's more recent history, but all are being given fresh life in recent Chinese military publications.

Finally, some distinctive aspects of China's contemporary air defense employment concepts appear to be a more archaic legacy of the general (world) history of aerial warfare, rather than of the Chinese experience per se. Attacks to "disrupt enemy formations," for example, seem to be based on the premise that enemy aircraft will be flying in close formation to provide mutually supporting cover fire, as the large bomber formations of World War II did, in contrast to the more-dispersed strike packages employed by modern air forces. And while the PLA's emphasis on the use of low-altitude approaches to achieve surprise may be based on its lack of stealth, its care to avoid falling prey to similar tactics on the part of the enemy appears to ignore the fact that many air forces—including the USAF—no longer rely heavily on low-altitude flight.

None of this is to say that Chinese air defense employment concepts, as they stand today, are not also a response to modern high-technology air war and China's interpretation of it. History and organizational culture shape the selection of China's responses to contemporary challenges, but the PLA is not a slave to either. Some historically successful concepts (e.g., antiaircraft ambushes) appear to Chinese air force thinkers to be well suited to today's technology and have been reemphasized. Others (emphasis on key-point defense and zone organization) do not and, although they remain distinctive legacy

---

[66] PLAAF, 1994, p. 135. The source does not say explicitly that the targets were U.S. aircraft, but the dates (1962–1972) and the fact that U.S. aircraft were destroyed over China during this period make them the most-likely targets.

[67] Cui et al., 2002, p. 333; Peng and Bi, 2001, pp. 84–91.

features of China's air defense employment concepts, are clearly being deemphasized and phased out in favor of more-adaptive operational concepts.

CHAPTER SEVEN

Air Blockade Campaigns

Air blockade campaigns are another air force campaign type discussed by the PLA. An authoritative source describes them as "offensive air combat implemented to cut off the enemy's traffic as well as economic and military links with the outside world." These campaigns are, according to this source, carried out "mainly by air forces, under the support and cooperation of other services and local armed forces." "An air blockade mission," the source continues, "is often carried out simultaneously with ground and maritime blockade missions; it is implemented separately only under special circumstances."[1] The scope of air blockade campaigns is therefore defined more by the lead service (i.e., the air force) than by the targets against which they are directed; they can be directed against maritime and ground transportation links, as well as aerial links with the outside world.

Air blockade campaigns first appeared in Chinese military writing in the late 1990s and in official doctrinal statements in 2000.[2] Despite being included in lists of basic air force campaign types, discussion of them differs from that of the other three. Air blockades constitute a narrower category than offensive or defensive campaigns. In many important ways, they can be regarded as a subcategory of offensive campaigns. They are clearly offensively oriented, and the organization and disposition of forces is similar to that of other offensive campaigns. At the same time, air blockades represent a larger collection of different

---

[1] Bi, 2002, p. 356.

[2] Wang Houqing and Zhang Xingye, 2000, pp. 363–365.

types of military efforts than the fourth type of air force campaign, an airborne campaign.

Air blockades could clearly be employed against Taiwan. Certainly, the timing of their appearance in the literature is more than coincidental. But, while Taiwan was almost certainly a key factor in the development of this concept, the literature on these campaigns also draws heavily on the lessons of recent U.S. air campaigns, including the Kosovo operation, battlefield air interdiction during the first Gulf War, and the implementation of the no-fly zones in northern and southern Iraq. The PLA discussion of air blockade campaigns adds to the more general Chinese discussion on the political uses of airpower for coercive purposes. They are, according to Chinese writers, fundamentally political operations requiring particularly careful political control but, if well executed, offering the promise of significant gains with potentially limited costs.

Air blockades can be conducted by the air force or, more usually, as part of joint campaigns that also include army, navy, and/or Second Artillery components. According to the *Campaign Theory Study Guide*, the purpose of the air blockade is to

> force the enemy into submission by carrying out a series of air blockade operations to cut off the enemy's external traffic, constrain the enemy's economic and military links with the outside world, deplete the enemy's economic resources, and weaken the enemy's potential battle capabilities.[3]

Although air blockade campaigns are, according to this source, a type of offensive air campaign, they have distinctive characteristics. Under normal circumstances, the objective of these campaigns is not the "large-scale destruction of enemy forces." Its emphasis, rather, is on "prohibition" [禁]. The *Campaign Theory Study Guide*, however, notes,

> under modern high-tech conditions when forces are locked in a fierce struggle with the enemy, attacks against enemy "counter-blockade systems" (enemy bases) will be necessary. Hence, in a

---

[3]　Bi, 2002, p. 356.

majority of cases, air blockade campaigns require a "prohibition-strike combination" [禁打结合].[4]

Evolution of Chinese Thinking on Air Blockade Campaigns

Perhaps nowhere is the fluid nature of Chinese thinking about air-power employment concepts more in evidence than in the discussion of air blockade campaigns. Originally discussed as an air "tactic," air blockade campaigns were elevated to the status of basic campaign type between 1996 and 2000. The recent discussion of these campaigns includes disparate elements of older and newer strands of thought—tactical and strategic—and continues to evolve.

The 1996 *Air Force Dictionary* listed only three fundamental types of air campaigns: offensive air campaigns, air defense campaigns, and airborne campaigns.[5] As of 1996, then, the PLAAF did not recognize "air blockade" as a separate campaign form, although the 1996 dictionary does include an entry for "air blockade" in its section on air force tactics.[6] The discussion of air blockades is limited in scope and centers on the blockade of airfields and transportation networks and the isolation of surrounded enemy forces. Firepower and air battle are the primary means discussed to achieve results. As of 1996, there appeared to be little sense that air blockades could be primarily political, that they could constitute military actions short of war, or that they could employ no-fly zones as important measures.

Four years later, when the 2000 edition of *Study of Campaigns* appeared, "air blockade campaigns" made their appearance as one of three air force campaign types (together with offensive and defensive

---

[4]   Bi, 2002, p. 356. This discussion of objectives and general methods is similar to that found in PLAAF, 1994, p. 102.

[5]   This source, like the later *China Air Force Encyclopedia* (PLAAF, 2005), provides English translations for key terms. *Zhanyi*, which is translated as *campaign* in later sources, is translated by the 1996 dictionary as *operation* (*Air Force Dictionary*, 1996, p. 17).

[6]   *Air Force Dictionary*, 1996, p. 29.

air campaigns).[7] In this source, blockade campaigns are regarded as "a type of strategic operation, the success or failure of which is connected to the highest interests of the state."[8] They are, in other words, strategic in nature and potentially decisive. They also, according to this source, have a strong "political and policy nature." This represents a major break in thinking on blockade operations and, arguably, airpower in general.

Both the 2002 *Campaign Theory Study Guide* and the 2005 *China Air Force Encyclopedia* reflect a further refinement of Chinese thinking on this subject, introducing the idea of no-fly zones as an important method in executing air blockade campaigns.[9] In principle, these documents leave open the possibility that a wide range of different types of campaigns might be covered under this campaign category, and the historical examples cited include broad variation in scope, mission, and circumstances (see the next section). In the description of forces, sequences, and operations, however, no-fly zones and the activities necessary to establish and support them take center stage.

In the 2006 version of *Study of Campaigns*, the discussion of joint blockade operations includes language stating explicitly that these may constitute "military operations other than war."[10] The 2006 *Study of Campaigns*, however, also provides a minor mystery. While listing them among the four basic air force campaign types, "air blockade campaigns" receive only a single paragraph in the introductory chapter to air force campaigns (chapter 27) and do not receive a detailed subsequent chapter elaborating on such campaigns. The other three types of campaigns (offensive, defensive, and airborne) each receive detailed treatment in their own chapters (chapters 28, 29, and 30).[11] The significance of this omission is unclear.

---

[7] Wang Houqing and Zhang Xingye, 2000, p. 350. Airborne campaigns were listed as a type of joint campaign in this work.

[8] Wang Houqing and Zhang Xingye, 2000, p. 363.

[9] Bi, 2002; PLAAF, 2005.

[10] Zhang Yuliang, 2006, p. 293.

[11] Zhang Yuliang, 2006, p. 558.

One possibility is that the omission signals a rethinking of the distinct nature of air blockades—at least as campaigns that would be run relatively independently by the air force. (Joint blockade campaigns receive an entire chapter, chapter 12.) The 2002 *Campaign Theory Study Guide* states, "air blockade campaigns belong to [the category] of offensive air campaigns."[12] Although this statement is followed up with a description of special features of the air blockade, it does raise the question of why this form of the offensive—and not others—is designated as a separate type of campaign.

A second possibility is that the sparse discussion of air blockade campaigns in the 2006 edition of *Study of Campaigns* has no particular significance. Air blockade campaigns were listed fourth among the campaign types in the earlier (2000) edition, while they are listed third in the introductory chapter on air force campaigns in the more recent (2006) edition (although, as noted, unlike the other three types of air force campaign, they do not have their own separate chapter).[13] In that sense, at least, it would seem that the air blockade has gained, rather than lost, standing.

A final possibility is that there is ongoing debate about the place of air blockade campaigns—a possibility that appears buttressed by the expanded discussion of challenges and obstacles facing such campaigns. This interpretation would suggest that air blockades will remain an important part of the PLAAF's repertoire but that a more nuanced view of the applicability and importance may be emerging. Our assessment is that, while this question is currently unanswerable, the second and third possibilities are more likely than the first. A new, more practical view of air blockade campaigns may be emerging, but these campaigns are likely to remain important in Chinese doctrinal thinking.

---

[12] Bi, 2002, p. 356.

[13] Zhang Yuliang, 2006, p. 558; Wang Houqing and Zhang Xingye, 2000, p. 350.

Historical Referents for Air Blockade Campaigns

The scope of air blockade campaigns varies tremendously and may include campaigns to isolate entire countries or specific sectors of the battlefield (often a city and its major transportation arteries). It is, then, particularly useful in this case to review the Chinese historical referents for these campaigns. The *China Air Force Encyclopedia* lists four.[14]

The first is the Soviet air blockade of German units surrounded at Stalingrad from November 1942 to February 1943. According to the encyclopedia, three air army groups, five airborne divisions, and 400 AAA guns were employed, with much of this force operating between Stalingrad and potential relief forces farther west. These forces destroyed 1,200 German aircraft, reduced German supplies to 17–25 percent of normal requirements, and sped the collapse of German resistance. This campaign provides perhaps the best example of the use of combined arms and the exploitation of positions behind enemy forces to isolate a part of the front.[15]

The second example is the U.S. Operation Starvation conducted against Japan during March–August 1945. This was an aerial mining campaign designed to shut down Japan's last links with its Southeast Asian empire by closing harbor entrances and sea routes. In all, 12,000 mines were laid, and 670 ships sunk. More importantly, ports and shipping were paralyzed, sea communications severed, and the collapse of Japan accelerated. This example is perhaps of most relevance for a Taiwan scenario, and aerial mine laying is regarded as one of the primary means employed in aerial blockades.[16]

PLAAF, 2005, suggests that the use of aerial blockade has received even greater emphasis since World War II. It has, according to this source, played a particularly significant role in limited war [局部战争], citing the coalition air blockade of Iraq during the 1991 Opera-

---

[14] PLAAF, 2005, p. 102.

[15] PLAAF, 2005, p. 102.

[16] PLAAF, 2005, p. 102. Mine laying is, according to the 2000 version of *Study of Campaigns*, one of four of the important operations conducted during air blockades (Wang Houqing and Zhang Xingye, 2000, pp. 365–366).

tion Desert Storm as the third example of an air blockade campaign, and NATO's blockade of Kosovo in 1999 as the last.[17] In addition, although not mentioned explicitly, much of the discussion of methods and implementation for air blockades, particularly the extensive discussion of no-fly zones [禁飞区], is reminiscent of the imposition and maintenance of no-fly zones in northern and southern Iraq by U.S., British, and French aircraft between 1991 and 2003.

Characteristics and Requirements

In addition to characteristics common to offensive air campaigns more generally, air blockade campaigns have several distinctive characteristics. The four official sources that discuss air force campaigns published since 2000 all agree on three such characteristics: They have a strong political and policy nature; they are of long duration; and their requisite operations are varied and command requirements are high.[18] Here, we summarize each of these three and then move on to discuss additional characteristics cited and discussed in Bi, 2002, and, especially, Zhang Yuliang, 2006.

Air blockade campaigns possess a strong political and policy nature [政策性]. As a campaign type with strategic military objectives, the ends are inherently political. Moreover, because blockades impinge on the interests of third countries (most importantly, the freedom of navigation of these states), they are governed by international laws. These laws include the London Declaration of 1909 that requires, among other things, that the geographic scope and effective dates of a blockade be established and declared in advance.[19] Even a slight neglect or violation of these laws may arouse criticism and "place the nation in a

---

[17] PLAAF, 2005, p. 102.

[18] Wang Houqing and Zhang Xingye, 2000, pp. 363–365; Bi, 2002, pp. 357–358; PLAAF, 2005, p. 102; Zhang Yuliang, 2006, pp. 292–294.

[19] PLAAF, 2005, p. 103; Declaration Concerning the Laws of Naval War, 208 Consol. T.S. 338, 1909.

reactive or weak political and diplomatic position."[20] Hence, there is a need for "close harmony" between the military effort and "national and political struggles."[21]

Air blockade campaigns are of long duration. Because the air blockade campaign does not seek a decisive battle with the enemy but rather seeks to deplete and exhaust the enemy through flight prohibitions, combat is likely to last for a prolonged period. To achieve the erosion of enemy strength, a large force must be committed, and these campaigns generally require some duration before the effects begin to show. Injunctions on blockade operations against islands, "especially large islands," are revelatory about Chinese thinking on their application to Taiwan operations. In these cases, if the forces for blockade are inadequate, given long distances and the number of possible targets, "we must adopt the method of local or partial blockade to achieve mission objectives."[22] In these cases, the campaign is likely to be of particularly long duration.

Requisite operations are varied, and command requirements are high. Air blockade campaigns involve noncombat operations, such as continuous monitoring of no-fly zones and inspection of unidentified aircraft, and they also involve combat operations, including intercepts and attacks on antiblockade forces. They include air force operations but also activities by other services.[23] Command and support are, therefore, highly complex and require a high degree of operational control (or initiative) and rapid adjustment.[24] To maintain the initiative, the blockading side must flexibly adjust various elements of the mission—including the targets and direction of the blockade, the extent and pace of operations, and the composition of component forces—in accordance with the mission and the enemy's reactions.

---

[20] Bi, 2002, pp. 357–358.

[21] Bi, 2002, p. 360.

[22] Bi, 2002, p. 358.

[23] PLAAF, 2005, p. 102.

[24] Bi, 2002, p. 358.

In addition to these three basic characteristics, some sources list others. *Campaign Theory Study Guide* includes a fourth characteristic. Air blockade campaigns, it stipulates, impose demanding requirements for combat conditions. A favorable international environment is mandatory, as "we can only achieve the best result when . . . we can win over understanding and support from the international community and most countries in the world."[25] In addition to the proper international conditions, blockade campaigns also require sufficient combat forces and operational support. Because these campaigns are generally of long duration and involve heavy combat losses, the scale of operations must match capabilities.

The 2006 edition of *Study of Campaigns* lists a total of six characteristics of joint blockade campaigns.[26] Although there is no separate discussion of air blockade campaigns, all the points listed for a joint blockade campaign pertain to an air blockade campaign. The characteristics cited by this source largely represent a finer and more detailed parsing of many of the points just made, but the overall tone and emphasis highlight the difficulties and hazards of blockade campaigns.[27] For example, this text reports that "battle area disadvantages are prominent and requirements for comprehensive support high" in a blockade campaign. Friendly forces will be projected forward to "high threat areas." Consequently, "successful execution requires reliable intelligence support, uninterrupted communications support, reliable

---

[25] Bi, 2002, p. 358.

[26] Zhang Yuliang, 2006.

[27] The six characteristics are as follows:

> [1] The campaign plan and activities are subject to a variety of limitations, and the political nature [of these campaigns] is strong. . . . [2] The operational intensity [of these campaigns] is relatively low, while the duration is long. . . . [3] The status [of these campaigns] as military actions other than war is prominent, and the complexity is growing. . . . [4] Offense and defense are linked, and the struggle for campaign initiative is intense. . . . [5] Possible sources of loss of balance or coordination are numerous, and command and coordination difficulties are large. . . . [6] Disadvantages of the battle area are prominent, and reliance on comprehensive support for campaign activities is large. (Zhang Yuliang, 2006, pp. 293–294)

sea and air support, solid engineering support, campaign cover support, search and rescue support, etc."[28]

Organization of Forces

Based on the duties to be assumed during an air blockade campaign, the mission forces will be divided into four elements (five including support forces, which are generally assigned to the other elements):[29]

- The no-fly enforcement force [空中禁飞兵力] is composed of fighters, long-range SAMs, and associated support units (including early warning, surveillance, and refueling aircraft). Its main tasks are implementing aerial surveillance and control and disposing of violating aircraft. The bulk of this force is usually deployed close to the front. (See the next section for a discussion of deployments and deployment regions.)
- The air strike force [空中突击兵力] consists of bombers, fighter-bombers, attack aircraft, conventional missile forces of the Second Artillery, army tactical missile forces, and army long-range artillery elements. Its main tasks include launching attacks on targets related to the enemy's antiblockade force, including air force elements and bases and SAMs. This force is usually deployed with some forward and the bulk further back.
- The air defense force [防空兵力] includes fighters, SAM forces, AAA elements, and support elements (including radar and communication forces). This force provides early warning and surveillance and ensures stability of the battle area by resisting enemy counterattacks. The bulk of this force is deployed forward, with some deployed in rear areas.
- The campaign reserve force [保障预备队] includes elements of all of the aircraft types associated with the first three elements. Its

---

[28] Zhang Yuliang, 2006, p. 294.

[29] This information is from Bi, 2002, pp. 363–365; Lu, 2004, pp. 269–270; and PLAAF, 2005, pp. 102–103.

main task is to strengthen efforts in the primary blockade direction. Campaign reserves are located in rear areas.

Forces will be organized into fully functional combat groups that will be deployed along particular combat directions (sectors) and will include elements of each of these forces.

Deployment and Disposition

The elements outlined in the previous section are deployed in various directions, defined by geographic sectors and differentiated by primary and secondary efforts. Forces deployed in the primary blockade direction will be deployed in depth and should be capable of engaging in both prohibition and strike operations. In addition to the "directions," whose boundaries run perpendicular to the enemy, the blockade forces are deployed in three lines or regions, defined by distance or depth from the target.[30]

- *First-line region* [一线地域]. Forces deployed here will normally include fighters, attack aircraft, and long-range SAM and AAA elements, as well as radar, surveillance, and other relevant support aircraft. During the initial stages of the campaign, forces in this region should be capable of monitoring and blockading tasks, as well as performing air defense. Long-range SAM elements should be capable of striking targets at their maximum range in the primary direction.
- *Second-line region* [二线地域]. Normally, forces deployed here include fighter-bombers, attack aircraft, and part of the fighter force, as well as early warning and command aircraft and aerial refueling aircraft. Fighter-bombers and attack aircraft deployed here should be able to perform strike missions "either directly or after being refueled at frontline air bases." Early warning and command aircraft, fighters, and reconnaissance aircraft should

---

[30] This discussion is from Bi, 2002, pp. 365–366.

be able to reach predefined aerial surveillance zones to carry out assigned tasks without refueling.

- *Third-line region* [三线地域]. Forces deployed here usually include bombers, transportation aircraft, and campaign reserve forces. Bombers should be able to reach designated target areas without refueling. Transportation aircraft should be capable of delivering necessary forces and supplies to areas required by the air blockade campaign without refueling.

The deployment of these forces will be adjusted during the course of the campaign, depending on battlefield conditions and demands. In organizing the deployment of forces to their operational areas, usually fighter elements and ground-based air defense (SAM and AAA) elements are deployed first. The radar, EW elements, and part of the operational support units will then deploy, completing the air defense component and providing cover for the forces that follow. Finally, bomber, attack aircraft, and transportation elements will move into place.[31]

Command Arrangements and Service Coordination

Command of air blockade campaigns is usually broken into two levels: a theater command [战区指挥机构] and subordinate campaign sector commands [战役方向指挥机构].[32] The theater commander implements the overall planning for the air blockade campaign, organizes the movement of campaign forces and initiation of the campaign, coordinates with the various campaign sector commands, and manages operations. The number and position of campaign direction commands depend on the demands of the campaign. The areas of responsibility of these sector commands should overlap those of neighboring commands and should extend to cover as much of the air surveillance zones and no-fly zones as possible.

---

[31] Information on deployment sequence is from Lu, 2004, p. 271.

[32] The material for this section relies on Lu, 2004, p. 270.

Although air blockade campaigns are usually joint campaigns engaging other services in support, the air force usually provides the bulk of the forces, and an air force officer usually directs the campaign. The air force commander for the campaign and the command headquarters should organize coordination between the PLAAF and other services, between the various campaign directions, and between the various branches within the PLAAF.

As in the other campaign types, coordination is highly reliant on procedural control. In conducting air strikes, the coordination between the PLAAF and Second Artillery is based on a temporal division, with the Second Artillery presumably striking first (though this is left unstated in the sources). The coordination of different types of air force elements depends on the situation. Fighter aviation takes the lead in organizing the enforcement of no-fly zones. The air strike force commander, on the other hand, coordinates air strikes. The coordination between fighter aviation and ground-based systems (SAM and AAA elements) generally is accomplished through airspace division.

Establishing No-Fly Zones and Aerial Surveillance Zones

The operations and procedures used in a blockade campaign will vary greatly, depending on circumstances. Typically, one or more no-fly zones are established, within or through which enemy aircraft, shipping, and ground transportation are not permitted.[33]

For each of these zones, the commander will establish one or (more often) several aerial surveillance zones [空中监控区]. (See Figure 7.1 for a schematic diagram of the zones employed in blockade campaigns.) Aerial surveillance zones are used by surveillance and early warning aircraft to observe enemy activities in or near the no-fly zone. They are also used as standby areas for fighters awaiting orders to respond

---

[33] Except where otherwise noted, the discussion of no-fly zones and aerial surveillance zones is from Bi, 2002, pp. 361–362. It is consistent with the discussion (and schematic) found in PLAAF, 2005, pp. 102–103. Note that maritime and ground transport are explicitly included in this list of prohibitions, despite the Chinese designation as no-fly zones.

Figure 7.1
Air Blockade Campaign Schematic

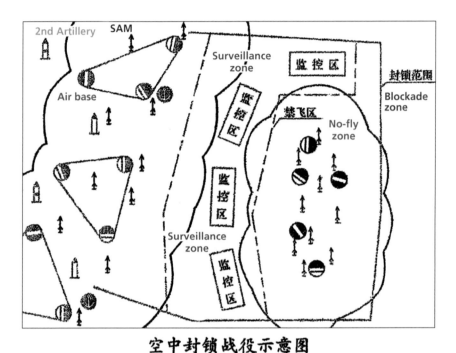

空中封锁战役示意图

SOURCE: PLAAF, 2005, pp. 102–103.
RAND MG915-7.1

to targets from enemy or third countries. Surveillance zones should be designed to meet the needs of the blockading force for conducting inspections of enemy or third-party aircraft or ships and expelling or executing forced landings against violators. Ideally, these zones should surround the no-fly zones they support, though the final designation of these areas will also depend on friendly and enemy capabilities.

Much of the writing on air blockade campaigns is devoted to the discussion of how no-fly zones and air surveillance zones should be chosen, as well as the more general question of the extent of the blockade zone. Determining the air blockade zone, the *China Air Force Encyclopedia* tells us, should depend on the political, economic, military, and diplomatic circumstances and should take into account international political and legal considerations.[34]

Operations Conducted as Part of the Air Blockade Campaign

A variety of different categories of operations will, according to Chinese military publications, be carried out as part of the overall air blockade campaign: information warfare operations, flight prohibition operations, interdicting enemy maritime and ground transportation, air strikes against the enemy counterblockade system, and air defense operations. Action plans for operations in each area should be established prior to the campaign's commencement.

Information Operations

In planning information operations, the forces available to both sides should be considered. The initiation of operations should be timed to catch the enemy completely off guard and combine information offense and firepower attacks on key information-related equipment or facilities of the enemy to achieve maximum effect. Equal attention is given to defensive information warfare. The mobility of equipment used in information warfare should be upgraded. Camouflage should be employed, and fake equipment should be shown and real equipment concealed. Finally, information warfare equipment should be improved so that it always operates in a secure mode.[35]

Flight Prohibition Operations

These operations are the primary combat operations of an air blockade campaign, and they will largely determine its success or failure. This category can itself be broken down further into five specific types of operations. *Aerial surveillance* is performed by aviation and SAM units in the surveillance zone to monitor enemy aircraft in preparation for

---

[34] PLAAF, 2005, p. 103.

[35] Bi, 2002, p. 368.

further action. *Aerial inspection* is carried out by aviation elements to identify and verify the identity of aircraft about to enter or traverse the no-fly zone. *Aerial expulsion* uses radio communication or other signals to warn off or expel enemy and neutral aircraft. *Forced landings* are performed by aviation units and involve forcing nonmilitary aircraft or third-country aircraft that have entered the no-fly zone or aerial surveillance zone to land at designated airports. Finally, *aerial attacks* employ firepower to destroy aircraft in the surveillance or no-fly zones that fail to heed warning notices or that attack friendly aircraft.[36]

Interdicting Maritime and Ground Traffic

In conjunction with the naval and ground force elements, air forces may also implement the blockade of maritime and ground traffic. Typically, maritime blockades are conducted jointly by the air force and navy and involve blockading maritime routes and attacks on shipping. Bombers and fighter-bombers are employed in blockading maritime routes, operations that generally involve mining port entrances and critical sea-lanes to impede and eventually sever transport traffic with the outside. Bombers, fighter-bombers, and attack aircraft may also conduct attacks on enemy merchant and military ships that have broken through the blockade.[37]

Strikes Against the Enemy Counterblockade System

Strikes on the enemy's counterblockade system can support blockades. Care should be used in choosing targets. Given that these campaigns are not designed to inflict the kind of massive damage that some other offensive air campaigns might, the primary targets should be those that pose a major threat to the blockade. These may include enemy air bases, air defense strong points, command and early warning systems, ammunition depots, and fuel storage. These strikes will generally be conducted by attack aircraft, covered by fighters, SAM forces, and other support elements. The first targets will usually include early warning systems and SAM systems, particularly those that are located close

---

[36] Bi, 2002, p. 369.

[37] Bi, 2002, pp. 369–370.

to Chinese positions but located on the enemy's periphery. Subsequent strikes may be conducted on air bases and related targets deeper in the enemy's territory. Attacks on air bases are generally led by conventional missiles of the Second Artillery, cruise missiles, and army tactical missiles, with aircraft conducting follow-up strikes. Attacks on the enemy's command and control should be led by information warfare forces and followed by vigorous attacks by combat forces. Methods for attacking ammunition and fuel storage will depend on their locations. Nearer targets should be attacked by aviation elements, supported by EW elements. More-distant targets should be attacked first by conventional missiles of the Second Artillery with follow-up by aviation elements.[38]

Air Defense Operations

Air defense is critical to the smooth implementation of various actions in the air blockade. In this case, there are three requirements. First, the force should perform full-time multidimensional surveillance and reconnaissance, combining space, aviation, maritime, and ground reconnaissance capabilities. Second, all organic antiaircraft power should be exploited. Ground-based air defense forces will serve as the core, but aircraft, missiles, and AAA should be coordinated under unified command. Third, forces should make maximum use of protection and camouflage. Mountainous terrain, vegetation, caves, and camouflage equipment can all be employed. Mobility will also enhance survivability.[39]

Concluding Thoughts: How an Air Blockade Campaign Might Be Implemented Against Taiwan

Air blockade campaigns are relatively new in the Chinese doctrinal repertoire, and the discussion of them is likely to continue to evolve. Here, it is worth considering how air blockades might be applied in future scenarios, particularly Taiwan-related scenarios. There is no

---

[38] Bi, 2002, p. 370.

[39] Bi, 2002, pp. 370–371.

doubt that the employment concepts for air blockades could be applied to (and therefore could support) a relatively comprehensive blockade of the main island of Taiwan. Historical examples cited in the Chinese literature include relatively large-scale efforts aimed at isolating entire nations (e.g., the U.S. Operation Starvation in 1945). Figure 7.1, for example, reproduced from the *China Air Force Encyclopedia*, suggests an extensive no-fly zone, one that includes five air bases and many SAM sites. Partial blockades of Taiwan would also be consistent with the concept, however, whether limited to a particular type of blockade (e.g., one executed primarily through mine laying) or to a particular portion of the island (e.g., the capital).

A blockade of one or more of Taiwan's smaller offshore islands would be more difficult to counter than a full or partial blockade of the main island of Taiwan. Moreover, several factors appear to make such employment of the air blockade concept more likely. From an operational standpoint, Chinese military publications have increasingly emphasized the difficulty of air blockade campaigns and the need to limit the campaign to areas where the implementing side can maintain an overall advantage of forces engaged, cover the surveillance zones with the engagement envelopes of SAM systems, and maintain a high operational tempo for extended periods of time. All of these conditions would be much easier to satisfy in an offshore island blockade than in a blockade of Taiwan Island itself. It would also be easier to satisfy the publications' injunctions against involving third parties in the conflict.

Finally, although any number of hypothetical routes to a China/Taiwan conflict might exist, a number of these might involve a decision to take concrete action of some kind to maintain Chinese credibility in the face of a perceived Taiwanese provocation. Based on past episodes of cross-strait tension and conflict, there could be significant differences of opinion in Beijing about the right course to take. Under such circumstances, a limited option that offered greater prospects for avoiding a full-scale conflagration but nevertheless satisfied the demand for action might look particularly appealing to Chinese political leaders. The point here is not that U.S. defense planners should not prepare to counter a large-scale air blockade of Taiwan, but that they should also consider how to respond to the operational and political challenges of more-limited air blockade campaigns.

CHAPTER EIGHT

Airborne Campaigns

Airborne campaigns [空降战役] are defined as combat action in the enemy depth carried out by airborne forces, air forces, and other forces of other services and branches. Usually, airborne campaigns are part of a larger joint campaign, but sometimes they are independent. They are extremely complex campaigns with many steps involved.[1]

PLA sources emphasize that one of the advantages of airborne campaigns is that they use long-range, surprise air raids to transcend natural, geographic, and man-made barriers. Airborne campaigns aim at the enemy's key points (transportation hubs, chokepoints, and centers of gravity) and at sabotaging the enemy's air defense system. In theory, they can even be used to achieve strategic goals, as the PLAAF notes that the Soviet Union achieved with airborne campaigns in Czechoslovakia (1968) and in Afghanistan (1979).[2]

Five principles arise repeatedly in PLA discussions of airborne campaigns. One, "full preparation and careful planning" [充分准备, 周密计划], includes planning and coordination between all services and forces involved in this complex campaign: airborne forces, air forces, and support forces. This principle also includes preparation for worst-case scenarios and alternative plans for the key actions in the campaign: air mobility, seizing information and air superiority, and conducting the ground offensive or defensive.[3]

---

[1] Zhang Yuliang, 2006, pp. 589–590.

[2] PLAAF, 2005, p. 103.

[3] Bi, 2002, p. 240.

A second principle, "concentrate forces and strike the enemy's pivotal points" [集中力量，打敌关节], includes striking against enemy command-and-control systems and transportation hubs—a common theme in PLA writings.[4]

A third principle, "win victory with stealth, suddenness, and surprise" [隐蔽突然，奇袭制胜], is also not unique to the airborne campaign; however, this principle is considered essential to the success of an airborne campaign because airborne forces attack deep behind enemy lines and enjoy relatively little support. To achieve surprise, PLA writings call for the use of feint attacks against false landing areas; concealing forces using darkness, weather, and electronic jamming; and landing close to the target to achieve objectives quickly and reduce risk.[5]

A fourth principle is maintaining "unified command and close coordination" [统一指挥，严密协同] between air and ground, airborne, and support and frontal forces during the actual campaign.[6]

Finally, a fifth principle is to "give prominence to focal points and strengthen support" [突出重点，强化保障]. Campaign commanders must organize adequate support so as to maintain air superiority, strengthen logistics, reinforce firepower, and strengthen reporting on weather conditions.[7] While these principles are common across PLA campaigns, airborne campaigns have their own particular challenges in implementing these principles.

According to PLAAF writings, an airborne campaign is resource-intensive and difficult: It requires air cover and defended air corridors for transport aircraft; it requires operations in which airborne forces can operate independently with little or no direct support on the ground; and it requires constant air cover, air supply, and air firepower support. Maintaining an effective command apparatus is also difficult: While the PLA is expected to "unify command and meticulously

---

[4]   Bi, 2002, pp. 240–241.

[5]   Bi, 2002, p. 241.

[6]   Bi, 2002, p. 241.

[7]   Zhang Yuliang, 2006, p. 593.

coordinate,"[8] it also notes several challenges to doing so: The battle-field is large; many different types of forces with disparate missions are involved (airborne forces, transport forces, attack aviation forces, reconnaissance aviation forces, electronic countermeasure forces, radar forces, and AAA forces, to name just a few), and command and coordination is subject to enemy interference. It requires strong communication support, electronic countermeasures, and an ability to distinguish friend and foe from the ground and in the air.[9]

Missions

Airborne campaigns' missions can include seizing enemy strategic points; seizing airfields, bases, and ports to facilitate landing operations; conducting sabotage, undermining the enemy's wartime potential or cutting off its forces; and undermining the enemy's command system and transportation hubs. The PLA probably would use airborne operations in a cross-strait conflict, either against the main island of Taiwan or against some of the smaller Taiwan-held islands in the Taiwan Strait area. ("Independently seizing enemy-held islands" is mentioned as a possible objective of an airborne campaign.[10]) Moreover, the PLAAF states that airborne operations are "an even more likely method to directly achieve strategic goals" in the future.[11] Potential objectives of an airborne campaign include the following:

- seizing and occupying enemy political, military, and economic centers or strategic points
- seizing and occupying important targets in strategic or campaign rear areas

---

[8] PLAAF, 2005, p. 103.

[9] PLAAF, 2005, p. 103.

[10] PLAAF, 2005, p. 103.

[11] PLAAF, 2005, p. 104.

- conducting sabotage and attacks behind enemy lines, damaging or destroying command centers, nuclear-weapon bases, and other important rear-area targets
- stopping the maneuvering and reinforcement of enemy campaign reserve units, destroying defense systems, or eliminating frontline enemy garrisons
- seizing and occupying critical targets in landing areas in coordination with frontal landing attacks. Such targets include naval and air bases, ports, and airfields.
- disrupting enemy command systems, communications, and transportation hubs.[12]

In addition to entire stand-alone "airborne campaigns," there are also "operational airborne landings" that are a component of other types of campaigns (such as an island-landing campaign). Operational airborne landings focus on achieving a specific operational objective, such as seizing and holding important targets in the enemy's rear, cutting off enemy campaign deployments, cooperating with a PLA frontline offensive or amphibious landing operation, and speeding campaign progress. Operational airborne landing objectives, methods, scale, location, depth, attack targets, and territory seized and held will depend on several variables: the enemy's combat strength and positioning, geography, climatic conditions, and the strength of airborne, transportation, helicopter, and support forces. A force usually consists of several regiments [团] or divisions [师]. A focused operation lacking robust support, airborne depth is usually no greater than 100 km, and operations generally do not last longer than three days and three nights. Forces can be delivered by parachute, helicopter, or aircraft.[13]

According to PLA sources, the advantage of parachute landings is that they are not very limited by terrain and are good for surprising the enemy. However, they generally drop over a large area, which can be problematic, particularly if heavy equipment is to be dropped. Landing in airplanes or helicopters is less risky but demands air and ground

---

[12] Bi, 2002, p. 239.

[13] PLAAF, 2005, p. 106.

cover, and airplanes require the availability of an airfield. While helicopters are less limited by terrain, they are not suitable for long-range, large-scale airborne combat. When parachute drops and air transport are combined, paratroops can land first, seize an airfield, and clear the way for transport aircraft to land.[14]

Planning Factors

PLA sources mention several parameters that could be regarded as planning factors for an airborne campaign: the selection of targets, departures, air transportation routes, landing sites, type of aircraft used, and timing of the operation. Preferred targets in an airborne campaign are identified as those that are both important and easy to attack, occupy, and defend (again, due to weak support for an airborne campaign). They should also be chosen so as to undermine the enemy's disposition and support and cooperate with the main front of the battlefield.[15]

Airborne landing sites should be close to these targets and should be in a location where the enemy's defenses are weak and there are no enemy forces (especially mechanized forces or tanks) or landing obstacles. The location should be 10–30 km from shore if airborne forces are going to assist landing forces in a landing campaign. If they are assisting attack forces on a land front, they should be dropped 30–50 km from the battle lines. Finally, forces should land in several different locations to shorten the time for landing. Of course, weather and terrain are also important planning factors for planning the location of a landing site.[16]

The best air transportation routes from friendly territory to the landing site take advantage of the enemy's blind spots, avoid dense firepower, and are a short and straight flight. If there are two air routes, they should be located about 30–40 km apart to be safe.

---

[14] Bi, 2002, p. 247.

[15] Bi, 2002, p. 245.

[16] Bi, 2002, pp. 246–247.

General Methods

In general, an airborne campaign should be covert and sudden, hit the enemy's key points, paralyze the enemy combat system and strike where the enemy is weak, and occur quickly. Airborne troops have little logistical support and a limited ability to fight a prolonged battle. Therefore, striking where the enemy is weak (avoiding tanks, mechanized force concentrations, or heavy defenses) and having brief battles that are quickly concluded are two particularly important concepts. Airborne forces simply are not well supported or equipped to sustain operations over a long period of time or engage in heated combat involving heavy forces.[17]

Composition and Deployment of Forces

The principal forces for airborne operations are drawn primarily from the air force, the army, and the Second Artillery. Typical operations incorporate paratroopers, air transportation units, fighter aircraft, bombers, reconnaissance forces, jamming aircraft [干扰航空兵], AAA, radar units, and SAMs. Larger-scale operations may also enlist the support of army infantry, other air forces, Second Artillery conventional missile forces, tactical missile units from other services, and reserve civil aviation units. Maritime-area operations may receive support from naval aviation and warships.

Air force units are responsible for a range of activities during airborne campaigns. PLAAF paratrooper [空降兵部队] objectives include seizing and occupying political, military, economic, or other strategic locations; landing in the enemy's rear area and opening new battlefronts; occupying and seizing transportation centers and key strongpoints to prohibit reserve forces from reinforcing enemy troops; assisting with the frontal assault; and seizing and occupying air and naval bases, ports, and other key targets. Aviation units [航空兵部队] are charged with conducting air reconnaissance and

---

[17] Bi, 2002, pp. 239–240.

counterreconnaissance; running electronic countermeasure operations; achieving information and air supremacy; administering preliminary air bombardments and opening up air corridors; and providing aviation fire support, air transportation, air cover, airborne supplies, and airborne command and control. Radar units, SAMs, and AAA units are responsible for airspace reconnaissance, guiding and directing air units' operational actions, resisting enemy countermeasures, and providing cover and air security for troop concentrations and important targets.[18]

PLA Army forces could also be involved in airborne campaigns. Air-landed troops [机降部队] can carry out ground attacks and defensive operations while air units conduct air strikes, provide air transportation, and enhance the firepower and mobility [机动能力] of landed troops. Army tactical missile units participate in the struggle to achieve air and information superiority, contribute to preliminary bombardments and the opening of air corridors, and support paratrooper operations.[19]

Second Artillery conventional missile forces serve a role similar to that of army tactical missile units. The Second Artillery is responsible for contributing its forces and capabilities toward achieving information superiority (presumably by attacking radars and other C4ISR installations) and air superiority, conducting preliminary bombardments, opening air corridors, and supporting paratrooper operations.[20]

Troops and equipment from the air force, army, and Second Artillery that participate in airborne campaigns are often organized into specific groups [集团] according to specific maneuvers [布势] associated with the campaign. These groups include the airborne combat group, the air transport group, the air cover group, the air strike group, the air support group, the missile strike group, the administrative support group, and the air defense group.[21]

---

[18] Bi, 2002, p. 242.

[19] Bi, 2002, p. 243.

[20] Bi, 2002, p. 243.

[21] Bi, 2002, p. 243.

- The airborne combat group [空降作战集团] consists mainly of paratroopers, infantry, and army aviation forces. The group's primary responsibility is to capture predetermined targets or to guard assigned areas. The group can be further organized into advance echelons, attack echelons, rear echelons, and follow-up echelons.[22]
- The air transport group [空中输送集团] consists of transport units and civil aviation reserve units. It is tasked with transporting airborne forces to the landing zone, delivering materials to the airborne combat group by air landing or airdrop, and transporting wounded soldiers. The group is deployed on the ground at second-line airfields or rear area airfields, depending on the aircraft's technical capabilities.[23]
- The air cover group [空中掩护集团] consists of fighter aviation forces. This group participates in seizing information and air superiority and provides cover and support for airborne, air transportation, and other groups. The group is deployed on the ground at frontline or second-line airfields; some are deployed at rear area airfields.
- The air strike group [空中突击集团] consists of bomber aviation forces, fighter-bomber aviation forces, and attack aviation forces. It participates in seizing information and air superiority, preliminary bombardments, and opening up air corridors. It also supports the ground combat actions of the airborne combat group. It is deployed on the ground at rear area airfields (bombers), second-line airfields (fighter-bombers), and frontline airfields (attack aircraft).
- The air support group [空中保障集团] consists of reconnaissance aviation, electronic jamming aviation, and early warning and command aircraft. It uses technical means to combat other aviation.
- The missile strike group [导弹突击集团] is composed primarily of Second Artillery conventional missile forces and army tactical

---

[22] Bi, 2002, p. 243.

[23] Bi, 2002, p. 244.

missile forces. Primary responsibilities for the missile strike group include helping to seize information and air superiority, participating in preliminary bombardments, opening air corridors, and supporting the ground combat operations of the airborne combat group. The group is deployed to operational depth or forward [浅近纵深] locations.

- The rear-area support group [后方保障集团] consists of logistics, offices for technical support of equipment, and various other support units. The group provides campaign logistics and equipment and technical support.
- The air defense group [对空防御集团] consists primarily of SAM forces, AAA forces, and radar units. It is tasked with conducting antiaircraft combat, destroying enemy attempts at preemptive combat and electronic countermeasures, and ensuring the air safety of the campaign's many combat groups. These forces coordinate with the air defense forces belonging to the army and navy to establish air defense in regional and strategic arrangements.[24]

Carrying Out the Campaign

Airborne campaigns incorporate several recommended "main combat methods," which emphasize maneuver, surprise, attacking key targets, attacking from unexpected angles, and seizing important targets in support of landing or attack forces, including dividing or splitting important areas in the enemy's rear areas.[25] Combat methods also include such moves as attacking from the side and from behind, seizing and occupying enemy areas in the campaign's forward areas or tactical rear; attacking enemy defense positions from the flanks to assist landing or attack forces; seizing and occupying airfields, landing areas, and other areas; and seizing and occupying enemy naval bases and ports in

---

[24] Bi, 2002, p. 244.

[25] Information in this section is derived from Wang Houqing and Zhang Xingye, 2000, pp. 479–484, and Bi, 2002, p. 249. Material in Wang Houqing and Zhang Xingye, 2000, is almost identical to that from Zhang Yuliang, 2006.

order to establish forward bases able to support landing troops going ashore.[26]

There are four main phases in an airborne campaign: seizing information and air superiority, carrying out firepower preparation, conducting airlift, and transitioning to the land battle.[27]

Seizing Information and Air Superiority

Seizing information and air superiority is viewed as key to the success of an airborne campaign. Information superiority is gained through information reconnaissance, attack, and defense. For an airborne campaign, air superiority is gained through several steps that differ somewhat from other campaigns due to the inclusion of air escorts, air blockades, and air patrols around the landing area:

1. Conduct firepower strikes and sabotage against important enemy air bases and enemy air defense warning systems, command-and-control systems, and air defense firepower systems.
2. Conduct air escorts and air blockades.
3. Conduct air patrols and area air defense (using AAA, SAMs, attack helicopters, and fixed-wing aviation).[28]

Personnel and heavy equipment are to be loaded and transported under the cover of darkness and weather. Boarding and loading should proceed according to a unified plan organized by joint commanders from the various airports from which payloads originate. Campaign commanders are responsible for overseeing [督促] close coordination among air operation groups, air transport groups, and transportation stations.[29]

---

[26] Bi, 2002, pp. 244–245.

[27] Bi, 2002, p. 245.

[28] Wang Houqing and Zhang Xingye, 2000, p. 480.

[29] Wang Houqing and Zhang Xingye, 2000, p. 480.

Carrying Out Firepower Preparation

Firepower preparation can be divided into two categories: advance firepower preparation and direct firepower preparation. In advance firepower preparation, the PLA destroys enemy ground force formations and support weaponry, suppresses and destroys the enemy's air defense system, weakens the enemy's counterairborne capabilities, and isolates airborne combat zones. Targets include enemy airfields, radar stations, ground-based air defenses, forces near the landing zone (especially tanks and mechanized forces), transportation hubs, and command and control, EW, and supply systems.[30]

In direct firepower preparation, forces attack and suppress the enemy's effective strength and air defense weaponry in the landing zone. They destroy defense facilities, interdict roads and bridges leading to landing zones, and attack the enemy's counterairborne reserve units (especially tanks and mechanized forces) and facilities.[31]

Conducting Airlift

During the third phase of the airborne campaign, airborne forces undertake a series of preparatory steps to open air corridors and conduct air transportation for airborne operations. This is the most complex and risky portion of an airborne campaign.

First, reconnaissance units report on the weather and the situation of the enemy. Jamming units jam the enemy's air defense system. Suppression units then destroy and suppress enemy radar, air defense missiles, and AAA positions. Finally, cover units destroy enemy aircraft. At this point, under cover from the cover units, transportation units fly airborne forces and equipment along designated air corridors to their landing zones. Early warning and command aircraft direct the actions of all groups involved. Before the transportation aircraft arrive at the beginning of an air route, cover units should first clear the airspace. Suppression units should conduct air blockades against enemy airfields and air defenses to ensure the safety of the transport

---

[30] Wang Houqing and Zhang Xingye, 2000, p. 481.

[31] Wang Houqing and Zhang Xingye, 2000, p. 481.

and airborne forces as they land.[32] The width of air corridors opened depends on the number of air transportation routes. Once an air corridor is opened, campaign commanders are responsible for ensuring that the enemy does not close it off.[33] Campaign commanders also are responsible for closely monitoring the enemy and changes in weather conditions, controlling the locations of various forces involved in this operation, coordinating their actions, and executing the landing plan unless major developments impede its execution.[34]

Transitioning to the Land Battle

In the fourth phase, the airborne campaign transitions from an air operation to a ground operation. During this phase, the airborne forces land and seize and establish an airborne landing base. Its successful completion lays the foundation for victory in ground operations. To be successful, the airborne landing group strives to maximize surprise and the effects of firepower strikes, minimize the duration of the landing operation, eliminate the enemy's presence at the airborne landing site, quickly expand the area of control, and set up a landing base for additional airborne landing operations.[35]

In order to seize and occupy an airborne landing base and guard strategic points and locations [夺占空降基地, 抢守要点要地], "advance echelons" [先遣梯队] of the airborne combat group must wipe out enemy forces at the airborne landing site, control tactical points in and near the landing area, set up and activate navigation equipment, report weather conditions, mark the location of the landing site, clear away barriers to landing, and direct and guide landing air formations. When landing on or near an airfield, these forces should quickly seize control of the airfield's command and communication facilities, clear away obstacles on the landing fields, and make sure that navigation equipment is set up to assist the landing of additional forces.

---

[32] Wang Houqing and Zhang Xingye, 2000, p. 483.

[33] See Zhang Yuliang, 2006, p. 597.

[34] Wang Houqing and Zhang Xingye, 2000, p. 483.

[35] Wang Houqing and Zhang Xingye, 2000, p. 483.

The next echelons to land are the "assault echelons" [突击梯队], which, after landing, seize and occupy strategic points and consolidate, expand, and occupy additional landing fields. If possible, they should link up airborne landing sites, repair any airfields now occupied, and construct temporary landing strips in the field of operations to ensure that rear and follow-on echelons can parachute or fly in quickly.

"Rear echelons" [后方梯队] land next. They quickly collect air-dropped combat weapons, construct a forward support base, and complete preparations for shipments in and out of the air base to ensure support to various units' operations.

Once the airborne landing base is set up, the "follow-on echelon" [后续梯队] of the airborne combat group parachutes in or lands. Occasionally, the airborne landing base is set up after completing the mission if the airborne landing group is tasked to carry out sabotage operations in the enemy's rear. Assuming that the airborne landing base is already set up, however, the follow-on echelon begins offensive or defensive operations immediately after landing. At the end of this phase, the campaign has transitioned into normal ground combat. At this time, the campaign commander should organize the main forces from the air strike group and the air cover group to support, cover, and protect these forces.[36]

Throughout the operation, the air cover group should be focused on providing cover for airborne landing operations. The main force of the air strike group, along with the missile strike group, should fire intensively against any enemy tanks, mechanized forces, or heavy forces maneuvering toward the airborne landing site. In addition, they should attack key routes and bridges and blockade (defensively) the airborne landing site.[37]

---

[36] Bi, 2002, pp. 249–250; Wang Houqing and Zhang Xingye, 2000, pp. 483–484.

[37] Zhang Yuliang, 2006, p. 601; Wang Houqing and Zhang Xingye, 2000, p. 484.

Concluding Thoughts on Airborne Campaigns

It appears that PLA strategic thinkers consider the airborne campaign to be an important operating concept that could have strategic effects (and has had, for other militaries). At present, it probably would play a supporting role in a PLA campaign and might be used for specific operational purposes. In an invasion of Taiwan, for example, an airborne campaign might be used to seize an airfield or a port, to seize key political or military command centers in attempt to paralyze or decapitate Taiwan's leadership, or to carry out sabotage against important enemy installations in their strategic rear areas. Moreover, writings on airborne campaigns have not changed almost at all in the past several years, probably indicating that little has changed in the PLAAF's operational approach to them. However, it appears that the PLAAF aspires to a more prominent role for airborne campaigns—or at least believes that they could have greater strategic importance for other militaries. According to the *China Air Force Encyclopedia*,

> following the development of military air transportation power, in the future airborne operations will be an important kind of campaign, and will be widely used. In local wars, airborne campaigns will be an even more likely method to directly achieve strategic goals.[38]

---

[38] PLAAF, 2005, pp. 103–104.

CHAPTER NINE

The Role of Other Services in Air Force Campaigns

The importance of conducting joint operations is a consistent theme of recent PLA publications, and the publications analyzed for this study frequently point out the roles that services besides the PLAAF would play in an air force campaign. Accordingly, this chapter provides a brief discussion of the role of the PLAN and Second Artillery in air force campaigns. The first section deals with naval air and air defense employment concepts. The second section deals with the Second Artillery's role in air force campaigns.

Naval Air and Air Defense Employment Concepts

An extensive treatment of naval air and air defense employment concepts is beyond the scope of this monograph, but we comment here on two of the most relevant aspects of PLAN employment concepts. First, we review Chinese writing on the defense of naval bases and naval air bases. Second, we briefly comment on Chinese fleet air defense concepts.[1] Third, we review Chinese concepts for air attacks on naval

---

[1] Note that the sources consulted for this section, including *China Naval Encyclopedia* 《中国海军百科全书》, Beijing: 海潮出版社 [Haichao Press], 1999, and Zhang Yuliang, 2006, had very little discussion of air defense for ships at sea. Since the present study was focused on materials relating to the PLAAF and its missions, it is possible that air defense of ships at sea is purely a PLAN mission and that significant PLA publications on this mission exist but simply were not found as part of our material-collection process. It is also possible, however, that the PLA does not yet have a well-developed doctrine for the air defense of ships at sea. Given the extensive amount of materials on PLAAF employment concepts we were able to

ships, as the PLAN is likely to take the lead in this area and little is said on the topic in PLAAF publications. Working from a smaller set of source material, the comments in this section are more tentative than other parts of this monograph.

Naval Base and Naval Air Base Defense Concepts

The defense of naval bases is part of the national air defense system.[2] Two material conditions give a distinctive flavor to naval air defense concepts when compared with those of the PLAAF. The first and most important is geographic. Located at the periphery of the country and closer to the enemy or enemy forces, naval bases are assumed to be under hostile observation. They often are subject to greater pressure from enemy EW. And because naval bases do not benefit from the strategic or operational depth that PLAAF bases tend to, they are under greater threat from sudden or surprise enemy attack.[3] The second distinctive feature derives from their constituent armament. Many of the navy's most effective air defense sensors and weapons are onboard its warships, and Chinese thinking on the defense of naval bases views warships as integral parts of the larger defense.[4]

The 2006 *Study of Campaigns* briefly discusses naval base air defense. Its discussion is largely consistent with writings on other aspects of national air defense (e.g., the defense of cities or larger theater air defense operations), except in its emphasis on the limited size of the battle space and the distinctive nature of the defensive weapons involved. It suggests that a portion of air defense assets should be assigned to the destruction of enemy support forces, such as AWACS, EW aircraft, and radar picket ships. Other assets should then be assigned to destroy enemy strike aircraft, with particular emphasis on the destruction of platforms (aircraft and ships) armed with guided or cruise missiles. In order to retain a measure of initiative, some ground-

---

acquire, moreover, and their lack of any discussion of air defense of ships at sea, at a minimum, it appears that this is not a PLAAF mission.

[2] See *China Naval Encyclopedia*, 1999, p. 477.

[3] Zhang Yuliang, 2006, p. 549; Cui et al., 2002, pp. 313–314.

[4] Zhang Yuliang, 2006, p. 553.

based assets (such as coastal or other SAM and AAA elements) should be positioned along likely routes of ingress to ambush incoming enemy aircraft, and then quickly repositioned to continue the battle. Finally, ships and aircraft should be on high alert, ready to disperse or intercept enemy aircraft or ships on notification of their approach.[5]

Cui et al. (2002) offers additional commentary on naval base air defense. Although this publication is somewhat less authoritative, the lead author is the deputy dean of the National Defense University and has experience in air defense, so his comments should be given some weight.[6] According to Cui et al., "guiding thought" [指导思想] on naval base air defense includes two major elements.[7] First, these operations should be joint operations, rather than independently conducted. Under high-tech conditions, "air defense [of naval bases] will be hard to achieve with only naval resources." This proposition reinforces the idea that PLAAF air defense assets will have to be "shared" with the other services, rather than the reverse. Joint air defense operations are said to require a unified plan for reconnaissance, warning, electronic countermeasures, and air defense firepower, as well as unified command of air defenses and means for sharing intelligence, the joint use of resources, layered defenses, and a mutually complementary system of air defense. But while there are theoretical methods to achieve this integration, it is apparently seldom practiced during peacetime.[8]

Second, resistance and counterattack operations must be combined, with even greater emphasis on counterattack than is true in PLAAF employment concepts. Unlike in the case of PLAAF air defense concepts, the counterattack is the "main" form of operation

---

[5]   Zhang Yuliang, 2006, pp. 552–553.

[6]   The lead author is PLAAF Major General Cui Changqi. The biographical information for him in the book says that he has "past participation in national air defense campaigns." The two other main authors are also PLAAF officers.

[7]   Note that, in both cases, the material in Cui et al., 2002, differs significantly from that found in Zhang Yuliang, 2006. The latter source says that naval base defense can be conducted independently by the navy or jointly (p. 547) and it says that, although "integrated attack and defense" is necessary, defense will be the main operation [以防为主，攻防结合] (p. 550).

[8]   Cui et al., 2002, p. 314.

in the defense of naval bases. This is likely a function of the lack of defensive depth enjoyed by naval bases, though writings on the subject also emphasize the importance of expanding the battlefield and achieving at least some depth. In counterattack operations, enemy long-range aircraft and aircraft carriers will be important targets but will be difficult to reach with aircraft alone. Therefore, planning for counterattacks should consider the use of Second Artillery assets against enemy bases and aircraft carriers, in addition to naval aviation strike assets.[9]

A variety of resistance countermeasures should also be employed in the defense of naval bases. Air defenses (SAMs and AAA systems) should be deployed in a multilayered, circular pattern, with medium- and long-range SAMs in the outer ring and short-range systems positioned inside for terminal defense. Camouflage and engineering will also be employed. Coastal contours can be used to hide surface ships, and engineering works can enhance survivability. Electronic countermeasures and deception should also be employed. Possibilities here include the creation of false targets and pretending that actual defenses have been destroyed.[10]

Fleet Air Defense

The *China Naval Encyclopedia* includes a surprisingly short section on fleet air defense.[11] It calls for the establishment of a circular, three-dimensional, layered system of defense. Resistance will be provided by firepower and electronic systems. Any fighters providing long-range air defense for the fleet will, under the direction of early warning aircraft, seek to intercept and destroy enemy aircraft before they can launch missiles. If the enemy succeeds in launching missiles, these will be destroyed by ship-mounted SAMs and close protection cannon sys-

---

[9]   Cui et al., 2002, pp. 314–315.

[10]   Cui et al., 2002, pp. 316–318.

[11]   *China Naval Encyclopedia*, 1999, p. 1589. A short paragraph (with no separate section heading) is provided on air defense within the section devoted to "defensive operations of surface ship formations." There is no discussion of air defense in the sections of the encyclopedia devoted to general operations or tactics—despite the fact that room was found for discussions of "ramming tactics" and "crossing the 'T.'"

tems. Long- and short-range decoys may create false targets, and ships will, when appropriate, undertake defensive maneuvers.[12]

Attacks on Surface Ships

The PLAAF sources analyzed for this study say little about strikes on naval units. Naval aviation would, in any case, take primary responsibility for these operations. The participating aviation forces are divided into three groups: an attack group, a support group, and a reserve group. The attack group is the main force executing the attack and consists of naval bomber aviation, fighter-bombers, and attack aircraft. Its primary responsibility is to attack the enemy ships. The support force consists of naval reconnaissance, fighters, fighter-bombers, and strike aircraft. Its task is to ensure that the attack unit performs its strike smoothly. The reserve force consists of the same types of aircraft as the attack unit and is assigned to deal with unforeseen situations or to expand the strike's effect.[13]

The main types of attack are concentrated attack, simultaneous attack, and sequential attack. Concentrated attacks involve multiple aircraft attacking a single important target in a formation (e.g., an aircraft carrier in a carrier battle group). Depending on circumstances, the attack may be launched from multiple directions or from a single direction. The support force may conduct reconnaissance, provide cover, and apply pressure or interference to support the attack force as it carries out its mission. Simultaneous attacks are used to attack multiple targets in a single formation. Finally, sequential attacks involve a relatively small aviation force attacking single or multiple targets over a period of time. Attackers in sequential attacks should vary ingress routes, altitudes, and methods of attack. Sequential attacks may be used following other types of attacks (e.g., concentrated attacks) to broaden results or increase damage. They may also be used to increase pressure over an extended period of time and disrupt or exhaust the enemy when a coup de grace is impractical. Or they may be used when

---

[12] *China Naval Encyclopedia*, 1999, p. 1589.

[13] *China Naval Encyclopedia*, 1999, pp. 442–443.

attacking ships with weak air defenses, such as supply ships or small naval vessels.[14]

In all cases, attacks should be tightly coordinated. In the future, *China Naval Encyclopedia* tells us, attacks by multiple aircraft types commanded and controlled by shipborne early warning aircraft will become the fundamental type of attack.[15] Today, however, coordination is far more likely to involve scripting and procedural control than more-dynamic methods. When cooperating with surface ships, aviation units usually attack first, and surface units follow up to increase the damage. Close coordination is imperative to avoid damage from friendly fire. When different types of aircraft platforms cooperate, those armed with antiship missiles generally attack first, followed up with attacks by torpedoes, bombs, rockets, and cannons to increase the damage.

The Second Artillery's Role in Air Campaigns

The Second Artillery participates in the opening strike phase of a conflict, including air force campaigns. It would most likely be quite active in an air offensive campaign or an airborne campaign. Second Artillery planning tasks include determining attack targets; determining the extent of damage to inflict (destructive, suppressive, or harassing attacks); choosing the type of guided missile to be used; determining the time of an attack; and drafting a guided-missile firepower plan.[16]

In general terms, the Second Artillery attacks military targets, transportation hubs, and economic infrastructure relevant to the enemy's military potential; targets with a support function for the enemy (ranging from depots to EW and command facilities); and targets that pose a direct threat to PLA forces (such as air bases). More-specific examples of these targets can include targets in the enemy's strategic rear, such as air force bases, naval bases, aircraft carrier battle groups,

---

[14] *China Naval Encyclopedia*, 1999, pp. 442–443.

[15] *China Naval Encyclopedia*, 1999, pp. 442–443.

[16] Bi, 2002, pp. 157–159.

and armed helicopter deployment areas; transportation hubs; supply centers (such as military machine shops, fuel depots, and ammunition dumps); heavy troop concentrations; EW platforms; and guided-missile, command, and warning systems.[17]

Attacks can be carried out as destructive attacks—concentrated firepower (missile) attacks to cause primary facilities to lose their operational capability. They can also be suppressive attacks—concentrated firepower combined with aircraft attacks to suppress enemy activity within a certain time period. Or they can be harassing attacks—attacks carried out at random that make it more difficult for the enemy to conduct regular combat missions.[18]

Guided missiles are selected based on the target that will be hit. For a hard, point target, a high-accuracy guided missile is used. To attack an area target, the guided missile does not have to be as accurate. When attacking enemy guided-missile positions, a blast fragmentation warhead or fléchette submunition warhead is recommended; if the missile is protected, then a penetrating submunition warhead is recommended. One can also use multiple warhead types at a time. For example, when attacking an airfield, one can use both a penetrating submunition warhead and a fléchette submunition warhead to damage the airfield and destroy ground-based aircraft.[19]

The following weapons and their uses are discussed in open-source literature:[20]

- weapons under testing
  - A fléchette submunition warhead [箭弹子母弹] attacks effective strength and weapons of the enemy exposed on the ground. These could include ground-based guided-missile launch equipment, aircraft on the ground, fuel depots, vehicles, air defense weapons, and ships anchored in harbors.

---

[17] Bi, 2002, pp. 157–159.

[18] Bi, 2002, pp. 157–158.

[19] Bi, 2002, pp. 158–159.

[20] Bi, 2002, p. 158.

- A penetrating submunition warhead [侵彻子母弹] is used against such targets as airfield runways, aircraft shelters, and semiunderground fuel facilities.
- Blast submunition warheads [爆破子母弹] attack targets that are relatively hard and resistant to attack. These could include railway stations, large-scale bridge spans, docks, harbors, semiunderground military infrastructure, fuel depots, ammunition dumps, and command centers.
- Weapons under research
  - a fuel-air explosive (FAE) warhead [云爆弹][21]
  - a terminal-sensing penetrating submunition [末敏侵彻随进爆破子母弹], used to attack airplane runways
  - a conventional antiradiation warhead [常规反辐射弹] that destroys electronic equipment.

---

[21] An FAE disperses a cloud of microdroplets of fuel into the air then ignites the cloud. They differ from conventional explosives in that a conventional explosive emanates from a single compact location. FAEs and other volumetric explosives emanate from a large volume of space and thus can have effects over a much larger area. Someone who is in a trench when an artillery shell explodes on the ground next to the trench will not get hurt so long as his or her head is down (though he or she might lose hearing). If an FAE is set off in the air near the trench, on the other hand, everyone in the trench will be flattened, due to overpressure from the resulting shockwave.

CHAPTER TEN

Possible PLAAF Operational Concepts, Capabilities, and Tactics in a Taiwan Strait Conflict

This chapter translates the general air force employment principles described in Chapters Four through Nine into a set of specific operational concepts and tactics for PLA air operations over and across the Taiwan Strait in the 2015–2020 time frame. The purpose of the analysis in this chapter is to identify the operational goals the PLA might seek to achieve and the operational concepts it might use in pursuing them. Key aspects examined here include how the PLA might protect its air bases and aircraft, how it might divide the airspace over and across the strait to facilitate and deconflict defensive and offensive operations by aircraft and surface forces (especially SAMs), and so on. Finally, we describe how the PLA might use the assets it is likely to have in the 2015–2020 time frame to achieve its goals.

Any exercise of this nature is inherently somewhat speculative. By its very nature, it goes beyond what is known from published documents. However, we have been careful to ensure that what we say in this chapter is consistent with what we have learned from PLA writings. We have also made an effort to highlight areas in which we are using assumptions, logic, predictions of PLAAF force structure and platform capabilities, or Western practices in similar situations to fill in gaps or extend our understanding of how the PLAAF might conduct operations against Taiwan in the 2015–2020 time frame.

This chapter is divided into four main sections. The first is a brief review of key aspects of Chinese military strategy and air force employment principles that are especially relevant to how future PLA air operations and tactics might evolve. The second logically applies these principles to how the PLA might deploy its air forces and protect them and other key assets in a conflict with the United States in the 2015–2020 time frame. The third section examines how the PLAAF might use its forces both to gain air superiority and to conduct offensive operations in a conflict with Taiwan and U.S. forces in the western Pacific. The final section summarizes key elements of the operational and tactical analysis and outlines implications for the USAF.

Review of Key Chinese Strategy and Airpower Employment Concepts

Most nations have unique "ways of war" that are derived from their cultural traditions, historical experience, political system, and economic capacity (among other variables). It has been observed, for example, that the current U.S. way of war tends to place great stress on casualty avoidance through the use of advanced technology and massive firepower. This section describes a Chinese way of war by reviewing key elements of Chinese military strategy and airpower employment concepts that have particular bearing on how the PLA might conduct air operations in the 2015–2020 time frame.

China's Military Strategy
Although China's military forces have made great gains in technical and operational capability during the past two decades and will continue to do so over the next decade or so, the overall balance of military capability in the 2015–2020 time frame will likely still favor the United States. Therefore, in any conflict in which China might face the United States as an adversary, such as in a conflict over Taiwan, Chinese strategists are faced with the challenge of defeating a militarily superior foe. An earlier RAND study found a considerable body of Chinese strategic writing that addresses the problem of how to defeat a

militarily superior foe.[1] This research identified seven prominent Chinese strategic principles for defeating such a foe. These principles are as follows:

1. Seize the initiative early in the conflict. This implies early offensive action against any forces of the superior power within reach.
2. Use surprise to amplify the effectiveness of initial attacks and assist in gaining and maintaining the initiative.
3. If possible, conduct preemptive attacks. This is closely related to principles 1 and 2. Initiating a conflict at a time and place of China's choosing maximizes the probability that it will enjoy an initial advantage in the local balance of forces and increases the chances of achieving surprise and seizing the initiative.
4. Rapidly raise the costs to the superior power. This principle attempts to leverage the perceived U.S. sensitivity to casualties and materiel losses to induce a state of collective shock and loss of popular will to persist in a conflict.
5. Limit strategic aims. Limited aims, if quickly achieved, hold the possibility of presenting the superior power with a fait accompli in which the cost of reversing Chinese gains exceeds the benefit the superior power is likely to attain.
6. Avoid direct force-on-force conflict and instead focus concentrated attacks at key weak points. In the case of a Chinese-U.S. conflict, this means that attacks against command systems, certain "low-density, high-demand" (LDHD) assets, and key support infrastructure and networks would figure more highly in Chinese planning than would direct engagements with U.S. combat forces.
7. Attack U.S. information and network systems in order to disrupt, delay, and confuse the overall U.S. response to Chinese actions. This principle could be applied via cyberattacks as well

---

[1]   See Roger Cliff, Mark Burles, Michael S. Chase, Derek Eaton, and Kevin L. Pollpeter, *Entering the Dragon's Lair: Chinese Antiaccess Strategies and Their Implications for the United States*, Santa Monica, Calif.: RAND Corporation, MG-524-AF, 2007.

as jamming of communications and even the destruction of key facilities or assets, such as command centers or satellites.

Taken together, these principles say that, if (for example) China chooses to initiate military action against Taiwan and the United States intervenes, or Beijing judges that the United States is *about* to intervene in the conflict, then the United States must be prepared for a series of attacks on key U.S. forces and facilities.[2] Such attacks would employ significant numbers of available Chinese combat assets and would be well planned and rehearsed. It is likely that they would be accompanied by massive cyberattacks on U.S. military and other government networks. The speed, reach, and increasing technical sophistication of China's air forces would make them crucial parts of such an operation.

Key Aspects of Chinese Air Force Employment Concepts
This section reviews certain key aspects of Chinese air force employment concepts described in earlier chapters of this monograph. The concepts of most interest here are those that allow us insight into specific operational or tactical approaches the PLA might take in a future conflict involving the United States.

Chinese military publications on joint campaign theory note several important principles for employing airpower. These include the following:

1. Take full advantage of airpower's rapid mobility, suddenness, and in-depth strike capability from the beginning of a campaign.
2. Concentrate the use of airpower against high-priority targets. This principle, in combination with strategic principle 6 in the previous section, is likely to result in initial surprise attacks against command centers, information infrastructure, support systems, and the like. Later in a conflict, we should expect to see focused efforts by the PLA to attack surviving or reinforcing LDHD assets, such as AWACS, Joint Surveillance Target

---

[2] This is not to say that such attacks would necessarily occur. China's leadership may well choose to refrain from attacking U.S. forces and facilities, or to limit the scope of such attacks, due to concerns about conflict escalation.

Attack Radar System (JSTARS), Global Hawk, bombers, and tankers, both in the air and on the ground.

3. Coordinate with other forces. This principle says that airpower should be used in conjunction with other forces to maximize overall joint effectiveness. In the context of a conflict over Taiwan, the key other-force elements with which China's air forces would need to coordinate would be the conventional SSM forces, SAM and other air defense units, naval combatants, and, later, if a landing occurred, PLA ground combat units requiring air support on Taiwan.

4. Protect the ability to conduct effective air operations. Special emphasis is placed on protecting aircraft, air bases, air defense sites, and radar installations.[3]

When taken together and combined with the seven military strategy principles discussed in the previous section, these four general principles suggest a good deal about the sorts of operations China's air forces would be called on to perform in a conflict with the United States over Taiwan. The first suggests that these forces would be very prominent in an opening round of attacks against U.S. bases and forces deployed in the western Pacific region. Principles 2 and 3 in *this* list suggest that air forces would coordinate their initial attacks with SSM forces to maximize their combined effectiveness against a range of targets identified as key to campaign success. For example, an initial wave of theater ballistic missiles (TBMs) might target ground-based air defense radar, command-and-control and missile sites, and the runways of military air bases. These attacks could be followed up later by ground-launched cruise missiles and ALCMs, as well as aircraft employing PGMs, against hardened aircraft shelters, aircraft in the open, and fuel-handling and maintenance facilities.

Throughout the conflict, the PLA would need to coordinate air defense measures between its SAM and fighter units both within and

---

[3] The principles listed here are based on three "guiding concepts" [指导思想] and six "fundamental principles" [基本原则] identified for air force operations in a joint campaign described in Bi, 2002, pp. 144–147.

across services. Both the PLAAF and the PLA Army operate land-based SAMs (and AAA), and now PLAN platforms have powerful air defense systems of their own, in addition to the PLAN's combat aircraft. Special attention would need to be paid to coordination of air defense measures for invasion transports and postinvasion resupply vessels in the event that an invasion of Taiwan took place. Key issues here would be to ensure the safety of surface vessels while minimizing the probability of PLAN escorts engaging PLAAF fighters or ground-attack aircraft supporting the invasion.

An additional concern that would persist throughout the conflict relates to the fourth principle for employing airpower: how to effectively defend PLAAF airfields, aircraft, and other key facilities from air attack by U.S. or Taiwanese forces.

Possible Air Defense Concepts and Methods

This section analyzes how China would likely conduct air defense operations in a conflict with the United States in the 2015–2020 time frame.

Procedural Deconfliction and Airspace Control

The preceding section noted the need for coordination between air force operations and those of other Chinese force elements. Before we examine specific Chinese methods and trends for air defense, it is worth reviewing how Chinese employment concepts envision deconflicting and controlling air defense (and, as we will see, air offensive) operations.

Chinese writings state that the primary means of achieving this coordination will be through the use of procedural controls. Procedural airspace control divides the airspace in question into different zones based on geographic location, altitude, and time. This allows the planners, or controllers, to allocate different "blocks" of airspace to different types of aircraft, missiles, or other platforms so that they can perform their assigned missions. For example, a fighter unit might be tasked to have four aircraft fly along a particular route, within a certain altitude range, during a certain time, to "sweep" for enemy aircraft. Another example might be the assignment of an AWACS unit to keep an aircraft continuously airborne over a certain point at a specific altitude.

Procedural airspace control is not the only possible means of coordinating and deconflicting air operations. An alternative is positive control. Under this scheme, controllers have a real-time picture of all friendly (and, ideally, all enemy) air activity and dynamically task, retask, and relocate friendly air and air defense assets as conditions demand. Compared to positive control, procedural deconfliction has the advantages of being simpler to implement, more robust in the face of enemy electronic or network interference, and requiring smaller command-and-control staffs, communication networks, and facilities. Its disadvantage is that it is generally less effective at extracting maximum combat power from available forces due to suboptimal and relatively inflexible space, route, and time allocations.

Most airspace control schemes are a blend of procedural and positive deconfliction and control processes, and the Chinese system in a future conflict would almost certainly blend elements of both approaches.[4] However, it is interesting to note that Chinese military publications place the emphasis on procedural control. This indicates a willingness to trade some combat effectiveness at the individual mission and platform level for a simpler deconfliction scheme that is less reliant on information networks and more robust to enemy electronic jamming and interference.

NATO forces took a very similar approach when confronting the powerful Warsaw Pact forces in the Central Region during the 1980s. However, since Operation Desert Storm in 1991, the United States has increasingly relied on positive control and deconfliction schemes in order to maximize both the efficiency and the effectiveness of its air efforts. This trend has been fostered by rapid advances in wireless and information network technologies on the one hand, and, on the other,

---

[4] For example, they would almost certainly use both adherence to airspace control procedures *and* proper IFF transponder responses to identify friendly aircraft.

a lack of sophistication and resources to effectively attack or disrupt these networks on the part of our adversaries.

Objectives and General Methods

As previous chapters have pointed out, Chinese military publications on air defense stress three overarching objectives of an air defense campaign and three general methods for achieving those objectives. The three objectives are as follows:

1. Protect the capital and senior leadership targets from air attack.
2. Protect important targets within the theater from air attack. These can include (but are not limited to) air bases, air defense radar and missile sites, military bases and fielded forces, command-and-control sites, population centers, and nuclear forces.
3. Seize and maintain air superiority.

These three objectives are pursued by implementing three general defensive methods or principles. The first of these is to intercept attacking forces as far as possible from their intended targets. The second is to present attackers with a deep defensive array containing both fighters and ground-based air defense systems. The third, and most recent, addition to Chinese air defense concepts is the notion of using limited offensive strikes against enemy air bases as a means of defense—this is analogous to the concept of counterattack in land warfare. In addition, it should be noted that there is to be a unified command of all air defense activities within each air defense theater or region.[5]

The three principles are implemented through the establishment of three air defense zones. The first of these is the interception zone. This is established far forward and extends to the forward edge of the theater or as far forward as Chinese fighters can effectively operate.

It is important to note that some, or all, of this zone might be beyond ground-based, or even AWACS, radar coverage, so that PLA fighters operating in this area would be forced to operate indepen-

---

5 See Chapter Six.

dently of ground control using their own radar, infrared search and track systems (IRSTSs), and other sensors to find, track, identify, and attack targets. This will make them both less effective and less efficient. Therefore, it is likely that, rather than mounting standing patrols, or even conducting frequent sweeps of this zone, the PLA would only occasionally sweep fighters through this zone with the goal of disrupting U.S. formations; harassing or destroying high-value assets, such as tankers, AWACS, JSTARS, or Global Hawk; and generally forcing U.S. forces to tie up resources in monitoring and defending a large area that could otherwise be used to facilitate offensive operations.

The next zone intruding U.S. aircraft or formations would encounter is the blocking and destruction zone. This area would present attacking U.S. aircraft with the full array of air defense threats. It would contain early warning and ground control intercept (GCI) radars able to detect intruders and vector interceptors toward them at very long ranges (200 nm or more from the radar site).[6] In addition, it would contain long-range SAM systems with their own acquisition and target-tracking radars and numerous missiles. These would be backed by a number of fighter orbits (known as defensive counterair combat air patrols, or DCA CAPs, in Western terminology). The forces in this zone would be tasked with destroying or disrupting most or all attacking formations. Figure 10.1 shows how the interception zone and blocking and destruction zone might be constituted along a stretch of the Chinese coast. Note that this figure does not come directly from any of the Chinese military publications examined for this study but rather represents an interpretation of how their concepts would be applied in the specific case of a conflict with the United States.

The third zone that any surviving U.S. aircraft would encounter is the deep covering zone. This zone would be defended primarily by shorter-range surface-to-air defensive systems situated to protect certain high-value targets. However, some fighters may also be assigned to patrol this area. It is also here that we can expect to find supporting assets, such as AWACS, tankers, EW aircraft and their bases. In addi-

---

[6] The depiction of the blocking and destruction zone in Figure 10.1 assumes that these facilities are located 50–100 nm inland.

Figure 10.1
Interception, Blocking and Destruction, and Deep Covering Zones

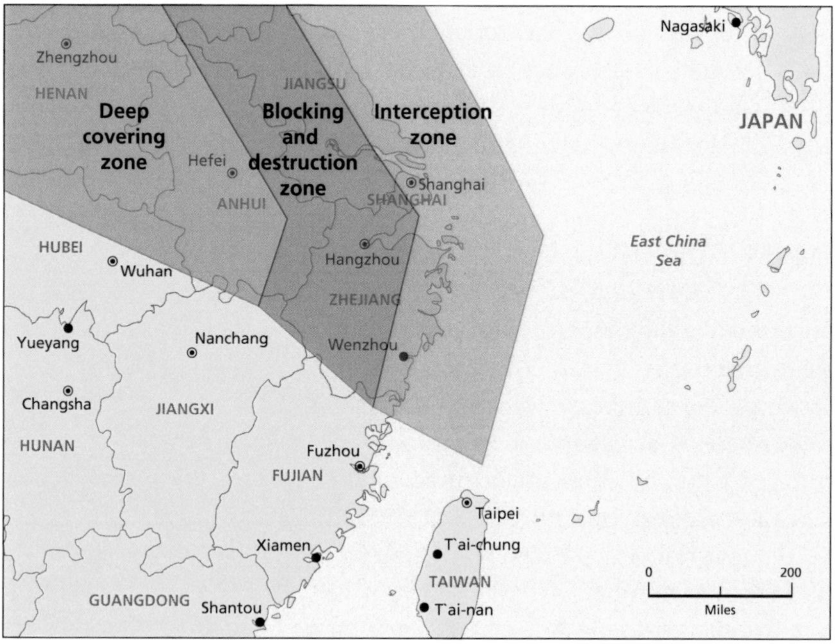

Zhengzhou
HENAN
Deep covering zone
JIANGSU
Blocking and destruction zone
Interception zone
Nagasaki
JAPAN
Hefei
ANHUI
Shanghai
SHANGHAI
HUBEI
Wuhan
Hangzhou
East China Sea
ZHEJIANG
Yueyang
Nanchang
Wenzhou
Changsha
JIANGXI
HUNAN
Fuzhou
FUJIAN
Taipei
Xiamen
Tai-chung
TAIWAN
0   200
GUANGDONG   Shantou
Tai-nan
Miles

tion, counterattack forces with the mission of attacking and destroying enemy air assets on the ground would be based here. Figure 10.2 depicts how the three zones might be implemented.

The depiction in Figure 10.2 is consistent with the Chinese concept of procedural deconfliction and control. The interception zone would be an area of fighter operations. Closer to the Chinese coast would be a SAM engagement zone, where friendly aircraft would be required to adhere to specific routes and procedures, and any aircraft not following safe transit procedures would be presumed hostile and engaged. The forward edge of this zone would coincide with the forward edge of the blocking and destruction zone. Behind the SAM engagement zone, but still within the blocking and destruction zone, would be an area with several standing fighter CAPs. These fighters would be there to protect the high-value assets located behind them

Figure 10.2
Notional Implementation of Chinese Air Defense Principles

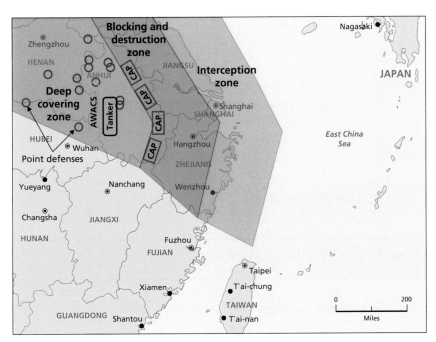

NOTE: SAM locations are purely notional.
RAND *MG915-10.2*

in the deep covering zone and to destroy any attackers that evade the SAMs. Finally, we see supporting assets, such as tankers and AWACS, along with small circles depicting terminal ground defenses of key points within the deep covering zone.

By the 2015 time frame, the fighter, radar, SAM, and network systems that constitute these defenses will likely consist largely of modern (though nonstealthy) fighters and advanced SAMs (such as the 150-km–range Chinese-made HQ-9, 100-km– to 200-km–range Russian-made S-300, the 400-km–range Russian-made S-400, and long-range early warning radars),[7] all connected by secure data and communication links. This will represent a significant change from

---

[7] Office of the Secretary of Defense, *Annual Report to Congress: Military Power of the People's Republic of China 2007*, Washington, D.C.: U.S. Department of Defense, 2007, p. 4.

a force dominated by aircraft and SAMs derived from the 1950s-era MiG-19 and SA-2 that made up the bulk of Chinese air defense capability until the late 1990s.

In addition to the defenses described in the preceding paragraphs and depicted in Figure 10.2, Chinese air defense concepts place a high priority on defending the capital of Beijing and the senior party leadership who live and work there. It is likely that significant air defense resources would be devoted to this task and that they would be among the best-trained and -equipped units.

Hardening, Camouflage, and Other Passive Defenses

Hardening, camouflage, and other passive defensive measures (such as having multiple aircraft runways and taxi-paths, ECM, and laser jamming systems) figure prominently in PLA air defense concepts. An examination of a typical PLA fighter base in east-central China—such as Fuzhou Air Base, depicted in Figure 10.3—illustrates both the seriousness with which the PLA takes this doctrinal point and the skill and resources devoted to implementing it.

Figure 10.3 clearly shows the main runway and, above it, the parallel taxiway, which could also serve as an alternative runway in the event that the main runway was disabled by U.S. attack. In addition, to the upper left of the parallel taxiway are two large, connected taxiway loops. Adjacent to these loops are more than 24 camouflaged, steel-reinforced concrete aircraft shelters. Figure 10.4 shows a close-up view of part of this area.

Four arch-type hardened shelters are visible at the lower left of Figure 10.4, with two more at the upper middle. These shelters are very similar to those built in Western Europe during the 1970s and 1980s to protect NATO combat aircraft. They are capable of protecting aircraft from the effects of almost any conventional weapon, as long as they are not directly hit. This would oblige an adversary to target each aircraft shelter with multiple PGMs to ensure that the shelter, along with any aircraft inside, was destroyed.

The extensive camouflage painted on the shelters and on the parking ramps and taxiways between them is also worth noting. While this would pose no challenge to a Global Positioning System (GPS)–guided

Figure 10.3
Overall View of a Typical Chinese Fighter Base: Fuzhou Air Base in East-Central China

SOURCE: Image courtesy of TerraServer. Used with permission.

RAND MG915-10.3

Figure 10.4
Camouflaged Hardened Aircraft Shelters at Fuzhou Air Base

SOURCE: Image courtesy of TerraServer. Used with permission.

RAND MG915-10.4

weapon, it has the potential to be effective in confusing other weapon-guidance techniques that rely on real-time human target identification for their success (for example, laser-guided bombs).

Aircraft shelters, such as those in Figure 10.4, are expensive to build—far more expensive than conventional hangar buildings. In addition, the need to disperse them and provide alternative access routes in the event that a particular taxiway is damaged by enemy attack requires a large amount of space. These last two factors limit the total number of shelters that can be built at any one base, with the result that fewer aircraft can be accommodated (if one wants to protect them all), and this increases the number of bases required to support a given number of fighter aircraft. The fact that the PLA is investing in these assets shows that it is "putting its money where its mouth is" when it comes to the airfield hardening, camouflage, and other passive defense techniques described in its publications.

Assessment

Overall, Chinese air defense concepts are quite sound. The PLAAF seems to be adapting to its acquisition of new, more-capable air defense systems (both SAMs and fighters). The best example of this is the adoption of the first-line interception zone concept. This acknowledges a need to keep enemy assets—especially intelligence, surveillance, and reconnaissance (ISR) platforms—as far as possible from Chinese airspace. It also takes advantage of the vastly increased ability of modern PLA fighters—such as the Su-27/J-11, Su-30, J-10, and FC-1—to operate more effectively outside friendly ground control coverage.[8]

However, the PLAAF has only a limited (though expanding) AWACS fleet, and this would likely limit radar coverage of the first-line interception zone to intermittent coverage (at best). Therefore, fighter sweeps through this zone would frequently have to rely on the fighters' own onboard sensors and luck to find enemy forces. It is possible, moreover, that U.S. aircraft operating in the zone would have good AWACS support, making such operations both less likely to make

---

[8] China uses the designator J-11 for Su-27 aircraft, and variants based on them, that are manufactured in China.

enemy contact and more likely to turn out badly for the PLA fighter force if contact is made, as compared to operations within the blocking and destruction zone.

Offensive Air Operations

As Chapter Five pointed out, the major tasks of an offensive air campaign according to Chinese military publications are as follows:

1. Conduct information warfare.
2. Penetrate enemy defenses.
3. Carry out air strikes.
4. Resist enemy air counterattacks.[9]

It is important to note that, if China chose a surprise preemptive attack, subsequent actions would be much easier to accomplish than they would be against an alert and prepared enemy. It is also important to recall that Chinese military writings emphasize the following general methods for conducting air operations:

- the use of surprise and preemption
- attacking a few vital points of the enemy system
- concentrating forces quickly and covertly
- using procedural deconfliction to coordinate offensive air efforts with defenses and other operations.[10]

Thus, just as with defensive operations, offensive operations would depend on procedural deconfliction and control. How these measures might be implemented over the course of an offensive air campaign against Taiwan and U.S. forces is described in detail later in this section. The next subsection focuses on how the PLA force structure in the 2015–2020 time frame could be used to support "opening moves"

---

[9] See Chapter Five.

[10] See Chapters Five and Six.

as part of an overall campaign plan for taking Taiwan by force in the face of U.S. intervention, based on the overarching concepts listed in "China's Military Strategy" and "Key Aspects of Chinese Air Force Employment Concepts" earlier in this chapter, with emphasis on gaining and maintaining the initiative through the use of preemption, surprise, and targeting of key enemy assets and facilities.

Long-Range Theater Strikes

The force employment concepts described in Chapters Four through Eight imply the need for capabilities that will allow China to attack and influence targets, enemy forces, and events not only on Taiwan but in more-distant areas of the western Pacific out to what is sometimes referred to as the *second island chain*. Figure 10.5 shows the geographic boundaries of both the first and second island chains.

Targets within the second island chain include the main Japanese home islands, the Marianas, and the Philippines. While the PLA has traditionally been primarily concerned with air defense and strategic deterrence, it appears to be in the process of acquiring significant new conventional attack capabilities that will allow it to effectively attack and influence enemy forces up to 2,300 nm (or more) from the Chinese mainland.

Foremost among these are new production variants of the H-6 bomber aircraft. The H-6 is a Chinese license copy of the Soviet Tu-16 Badger bomber designed in the early 1950s, which China has been producing since the 1960s. Early variants were used as free-fall nuclear and conventional bombers. More recently, the H-6 has been modified by both the Soviets and China to carry antiship cruise missiles and to conduct electronic reconnaissance and jamming, aerial refueling, and other roles. Of greatest interest in the context of possible preemptive attacks against distant enemy forces and bases are the most-recent developments of the H-6 line: the H-6M and H-6K.[11]

---

[11] For more on recent H-6 developments, see "Xian Aircraft Industries Group: XAC H-6," *Jane's All the World's Aircraft*, August 2, 2007, and especially Carlo Kopp, *XAC (Xian) H-6 Badger*, Air Power Australia technical report APA-TR-2007-0705, July 2007.

Figure 10.5
Geographic Boundaries of First and Second Island Chains

SOURCE: Excerpt from Office of the Secretary of Defense, 2006.
RAND *MG915-10.5*

The H-6M may be either a new production aircraft or a modification of an existing airframe with relatively few flight hours. In order to reduce aerodynamic drag, it dispenses with the 1950s-era gun turrets, gunners' windows, and various older radomes and antennas. It also has modern avionics, such as GPS and inertial navigation systems, modern radar, radios, and defensive systems. The most-significant changes include the deletion of the bomb bay doors and the permanent installation of fuel tanks (originally developed for the tanker variant) in the bomb bay and the installation of two additional wing pylons, for a total of four. These pylons can carry a wide range of antiship and land-attack cruise missiles, including the Chinese YJ-63 and DH-10 missiles.[12]

In either case, the 25 or so H-6Ms that are speculated to be currently on order would present quite a strategic asset for the PLAAF.[13] The aircraft themselves have an unrefueled combat radius of about 1,500 nm. Adding the range of the cruise missile to the combat radius of the bomber, it would be possible to accurately attack fixed targets more than 1,600 nm from the Chinese mainland using YJ-63 missiles or more than 2,300 nm using DH-10 cruise missiles. Assuming a mission-capable rate of 75 percent, it is quite feasible that a fleet of 25 aircraft of this type could put 19 aircraft and 76 cruise missiles into the air at one time.

The USAF has been investing heavily in new facilities at Andersen Air Force Base on Guam. These include climate-controlled maintenance facilities to support B-2 operations, as well as hangars for F-22s and other fighter aircraft. In any conflict with China, Andersen would be a key base for U.S. bombers, fighters, tankers, and ISR platforms (e.g., AWACS, JSTARS, Global Hawk, RC-135s). These would mostly be parked in the open or in unfortified hangers, as Andersen does not currently have hardened aircraft shelters.

The impact of 75-plus accurate cruise missiles—each with a warhead of 900 to 1,100 pounds—on the soft buildings and aircraft parked in the open would be devastating, especially if some of the weapons carried submunition warheads.[14]

Moreover, it seems that the H-6M may be only an interim solution to the PLAAF's theater strike capability requirement and that the

---

[12] Unclassified sources, such as SinoDefence.com, credit the Chinese H-6–derived tanker aircraft with a 3,250-nm (6,000-km) range. Since the H-6M is believed to have similar fuel capacity, this would make 1,500 nm a reasonable estimate of its operational radius, if we assume a 10-percent fuel reserve.

[13] See "Xian Aircraft Industries Group," 2007.

[14] For more on the likely effects of cruise missile attacks on aircraft in the open, see John Stillion and David T. Orletsky, *Airbase Vulnerability to Conventional Cruise-Missile and Ballistic-Missile Attacks: Technology, Scenarios, and U.S. Air Force Responses*, Santa Monica, Calif.: RAND Corporation, MR-1028-AF, 1999.

PLAAF is actively developing a new version of the H-6. This new H-6 version—the H-6K—is currently undergoing test.

The H-6K retains all the changes implemented in the H-6M and adds even more-radical modifications. The most obvious is the addition of two more underwing pylons (for a total of six). These are located near the wing roots. However, the most important modification is probably the larger engine intakes that are believed to lead to new turbofan engines that replace the thirsty 1950s-era RD-3M turbojets of earlier H-6 versions. The new engines (the D-30 turbofan used on the IL-76 transport and derivatives is often mentioned as a possible candidate in unclassified sources) are likely to have both more thrust and lower specific fuel consumption than the RD-3. This would allow the H-6K to take off at heavier gross weights (such as with full internal fuel and six missiles) and still achieve better range than the H-6M. Other important upgrades for the H-6K are an all-new "glass" cockpit with advanced avionic systems and the addition of ejection seats for the crew. Some sources also claim increased use of composite structures for reduced airframe weight.[15]

Figure 10.6 shows an H-6K's enlarged right engine intake. Another photo of an H-6K (not shown) clearly shows three cruise missiles suspended from the aircraft's right wing. If this aircraft enters production and replaces a large fraction of the 100 or so H-6 bombers currently in the PLA inventory over the next decade or so, the PLA will have a very formidable theater strike capability.

For example, if we assume that the existing fleet of H-6 bombers is replaced by a fleet of 25 H-6M and 75 H-6K over the next decade and we make the same 75-percent mission-capable rate assumption as we did for the H-6M example, then the PLAAF could deliver more than 400 cruise missiles to targets 1,500–2,300 nm (or more) from the Chinese mainland in a single mission. This sort of firepower, if concentrated on a few key targets, such as Andersen Air Base on Guam and

---

[15] Overall, H-6 range-payload and performance are similar to those of its contemporaries—the (now retired) British Victor, Valiant, and Vulcan bombers. For a detailed analysis of H-6K improvements, see "Xian Aircraft Industries Group," 2007.

Figure 10.6
Prototype Chinese H-6K Land-Attack Cruise Missile Carrier

SOURCE: "Air Power Australia," 2009.
RAND *MG915-10.6*

Misawa Air Base in Northern Japan, as an "opening move," as Chinese military publications stipulate, could be devastating.[16]

Even if it is not used, this capability could impose a substantial cost on U.S. efforts to defend Taiwan by forcing the United States to devote large numbers of high-end air defense systems (e.g., AWACS, F-22s, Patriot batteries) to guard against a cruise missile barrage like the one just described.[17] Figure 10.7 shows the likely extent of the H-6K strike radius from bases in east-central China.

---

[16] Air-launched cruise missiles could also be used against Kadena Air Base (and other air-fields) on Okinawa, but these are all within range of ground-launched cruise missiles and conventional ballistic missiles, so it is likely that these closer targets would be attacked by those systems while the air-launched cruise missiles would be used against longer-range targets.

[17] An analysis of how to protect a high-value target like Andersen from a massed cruise missile attack like the one described here would be an important first step in assessing this vulnerability. Among the most important aspects of such an analysis would be evaluating the effectiveness of existing ground-based air defense systems, such as Patriot, against large numbers of low-altitude targets, and devising an effective, robust, and affordable early warning system.

Figure 10.7
H-6K Cruise Missile Carrier-Aircraft Strike Radius from East-Central China

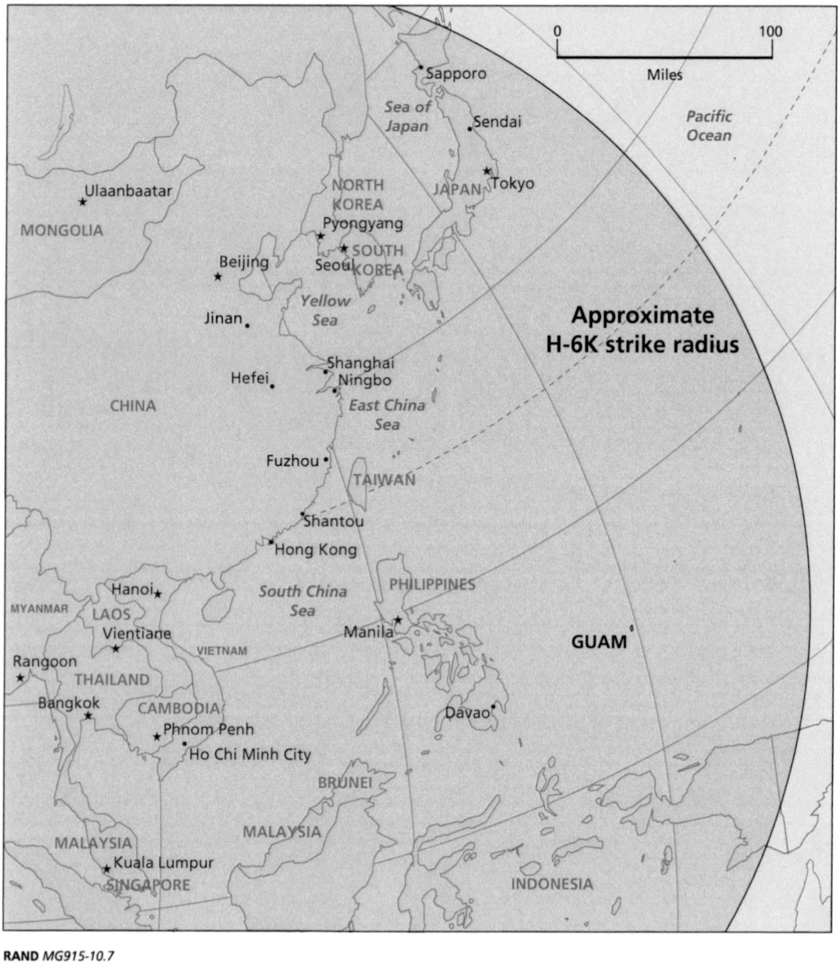

RAND *MG915-10.7*

The feasibility of such an operation would only be enhanced as the PLA acquires a number of Il-78 tankers that could allow modest numbers of Su-27 or Su-30 fighters to escort the H-6 fleet to distant targets.[18] While it may seem surprising that a fleet of modified 1950s-

---

[18] SinoDefence.com states that, in 2005, China ordered four Il-78s (based on the Il-76 transport aircraft) but that they have not yet been delivered.

era medium bombers could pose a serious threat in the 21st century, it is worth remembering that the USAF plans to operate its fleet of B-52s—aircraft very similar to the Badger in terms of technology and age—well into the 2030s in the same standoff missile carrier role.

The specific magnitude of the future threat posed by these new bomber variants is, of course, uncertain, but the PLA clearly seems to be in the process of acquiring capabilities that will challenge the assumption that Andersen and other distant theater bases in the western Pacific are immune to conventional attack.

Shorter-Range Offensive Operations

Closer to home—around the first island chain—future Chinese air force operations will be more intense and will face the requirement to integrate operations across weapon-system types (aircraft, SAMs, AAA, and SSMs) and across services (PLA Army, PLAN, PLAAF, and Second Artillery) far more closely than would be the case in distant theater strike operations. The elements that would require the most-detailed integration would be PLAAF and PLAN aviation offensive air operations with tactical ballistic missile strikes by the PLA Second Artillery (especially early in the war), as well as with long-range SAMs operated by the PLAAF on land and the PLAN afloat.

Possible Offensive Operations Against Taiwan

This section combines concepts from Chinese military writings with known characteristics of PLA systems to paint a plausible picture of how PLA offensive air operations might be conducted against Taiwan in the 2015–2020 time frame. As in the case of air defense operations, logic, common sense, and Western practices under similar circumstances are used to fill in gaps and areas of uncertainty in Chinese military writings.

As previously discussed, Chinese military writings place great emphasis on gaining and maintaining the initiative. Over the past 15 years, China has invested heavily in a large inventory of conventionally armed tactical ballistic missiles. These systems continue to grow in accuracy, numbers, warhead options, and range. These attributes, combined with extremely short flight times (usually under 10 minutes) make the tactical ballistic missile force operated by the PLA Second Artillery an ideal first-strike weapon for targets within its reach.[19]

For the most part, these will be targets in the first island chain. This includes targets on Taiwan and the U.S. air bases of Kadena and Futenma (or the future Futenma replacement facility) on Okinawa. Airfields, ground-based air defense sites, radar installations, and command-and-control facilities would be high on the list of targets attacked by the initial ballistic missile barrage. As the missile attacks drew down Taiwan's air defenses, it is likely that PLA fighters would move forward into offensive counterair (OCA) CAPs over the Taiwan Strait. Once these were established, they would likely be followed by integrated strike packages operating at low altitude against surviving Taiwanese air bases, air defense sites, and other high-value targets.

Initial Actions. As the long-range strike discussion made clear, the foundation of any successful military operation is secure bases. Therefore, it is likely that Chinese airspace opposite Taiwan would be defended by a combination of SAMs, fighters, early warning and GCI radars, AWACS, and other command-and-control systems similar to that described earlier in the section on air defense operations. However, in the event that China was conducting an air offensive campaign, it is likely that both the quantity and quality of these defensive systems would be higher in the area of offensive operations than in areas where only defensive operations were contemplated (with the exception of the Beijing region). The primary reason for this would be that China's best

---

[19] Over the next decade or two, it is possible that some of these systems will acquire even greater accuracy and possibly the ability to attack specific hardened aircraft shelters or even ships at sea. The most recent versions of the CSS-5 are already reported to be capable of extreme accuracy (50-m circular error probable, or CEP) or better over significant ranges (up to 2,500 km) due to its maneuvering reentry vehicle and terminal radar guidance system. See, for example, "DF-21 (CSS-5)," *Jane's Strategic Weapon Systems*, June 18, 2007.

air, naval, and army assets would be concentrated in the areas of the Chinese mainland opposite Taiwan and would provide tempting targets for preemptive action if not well defended.

The key procedural deconfliction challenge in areas where significant air offensive operations were planned would be to facilitate effective air attacks on Taiwan and perhaps U.S. bases on Okinawa and elsewhere while ensuring that Chinese bases and other key assets were well protected. At the heart of this challenge would be coordinating the efforts of PLA fighters and attack aircraft conducting offensive operations with PLAAF and PLAN long-range SAMs.

One simple (and therefore robust) approach to the airspace deconfliction and control problem is to rigorously segregate offensive and defensive operations in space and to further divide defensive sectors into areas defended by ground-based systems and areas defended by fighters. Figure 10.8 shows how the airspace near the Chinese coast opposite Taiwan might be divided prior to the start of offensive air operations.

Following the initial ballistic missile barrage, Chinese military publications suggest that the PLA would take several steps to secure the airspace between Taiwan and the mainland to facilitate rapid and effective follow-up air attacks. One way of achieving this would be to establish OCA CAPs close to the coast of Taiwan supported by AWACS and tanker aircraft operating behind the screen of advanced SAMs and fighters. Figure 10.9 shows the form this might take.

The area of AWACS coverage in Figure 10.9 assumes that the AWACS aircraft is operating at 30,000 feet and shows the area where such an aircraft could detect targets down to ground or sea level.[20] Targets flying at higher altitudes could potentially be detected farther away. Conversely, it might be possible for an aircraft on Taiwan to take advantage of terrain masking to approach one of the OCA CAPs at low altitude. Such a tactic would put the low-flying aircraft at a severe dis-

---

[20] The AWACS coverage area in Figure 10.9 is based on the maximum line of sight from an aircraft operating at 30,000 feet to the surface of the earth at sea level (assuming a spherical earth). It does not reflect the capabilities of any particular radar system.

Figure 10.8
Heavy Defenses Opposite Taiwan Prior to Initiation of Hostilities

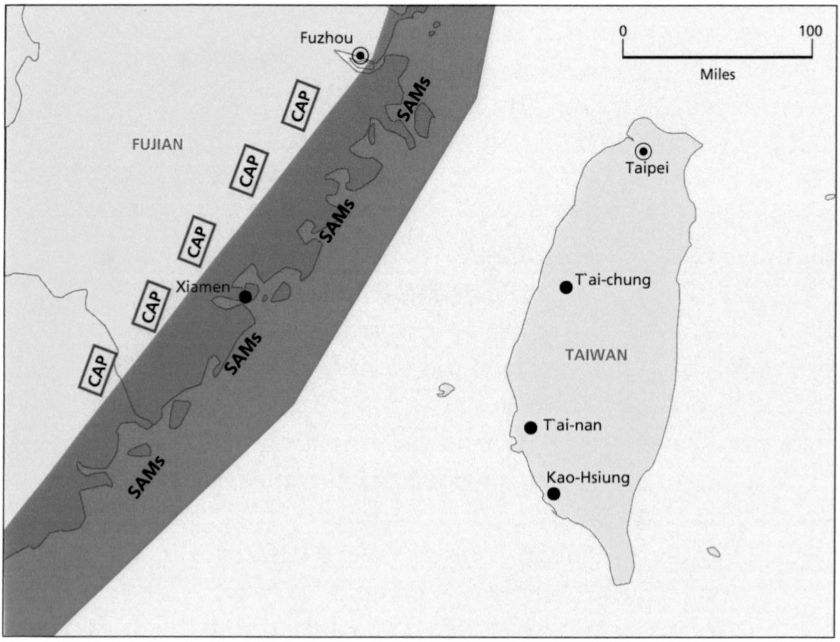

advantage relative to the OCA CAP fighters, however, as their altitude advantage would confer greater speed and weapon range.

Once the OCA CAPs were established, strike aircraft might begin to proceed down one of several air attack corridors. The Chinese military publications examined for this study are silent on such information as how many corridors might be created to attack Taiwan, what their dimensions might be, and where they would be located. Therefore, this section draws on standard Western practices to illustrate one way in which the PLA might implement an airspace control and deconfliction plan in an attack on Taiwan.

The first step would be to divide the airspace between the mainland and the Taiwan coast into several sectors. As Figure 10.10 shows, for purposes of illustration, we have chosen to divide the space into three sectors, each between 50 and 75 nm wide. Within each sector,

Figure 10.9
Controlling the Air Above the Taiwan Strait

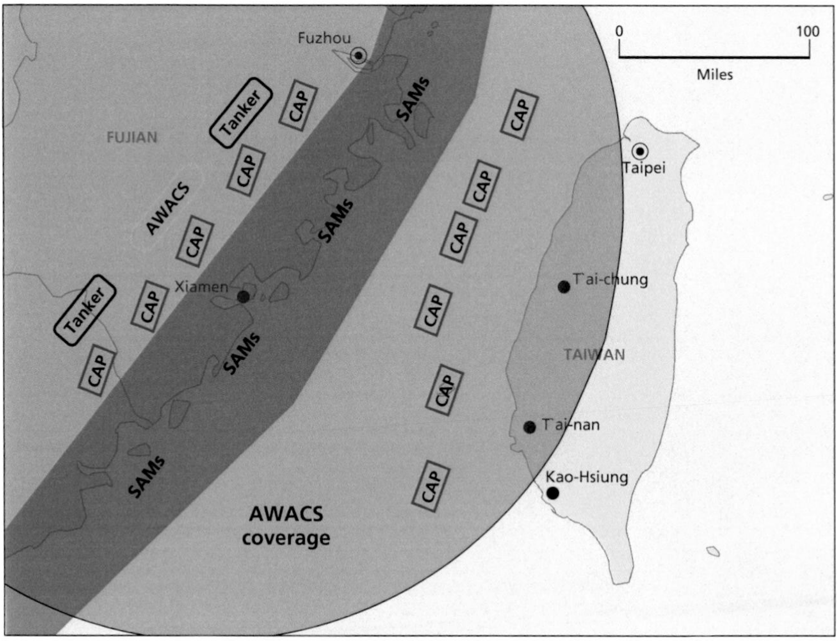

there would be a single pathway that all Chinese aircraft transiting to and from Taiwan would be required to follow. These pathways would be the air attack corridors. These corridors would likely be approximately 10 nm wide. Aircrews would be instructed to remain within the corridors and to fly at specific altitudes and airspeeds when in them. The goal of all of this would be to facilitate offensive air operations while, in conjunction with electronic IFF systems, minimizing fratricide.[21]

Chinese military publications, like U.S. and NATO doctrine of the late 1970s and 1980s, put great stress on low-altitude operations by large formations or "strike packages" of aircraft. Therefore, it is likely that outbound strike packages (those heading east to Taiwan) would be

---

[21] In reality, these corridors would probably have more-complicated configurations than depicted here, since, as described in Chapter Five, Chinese military publications emphasize attacking targets from multiple directions.

Figure 10.10
Possible Layout of Offensive Air Operations Sectors and Air Attack Corridors

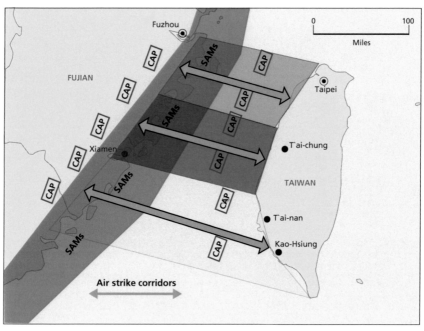

instructed to fly at very low altitudes and high speeds, while inbound aircraft (those flying west toward China from strike or CAP missions) would be required to fly somewhat higher and slower to present a less threatening picture to air defenses than would be expected of enemy attack aircraft.

Under a system of procedural airspace deconfliction and control, any aircraft flying in the airspace between China and Taiwan but not adhering to the corridor altitudes, speeds, and routes would be assumed to be hostile and would be subject to attack unless it was transmitting the correct IFF codes. The corridor routes would probably be moved several times each day (perhaps at seemingly random intervals) so that enemy aircraft would not be able to take advantage of them to "sneak into" Chinese airspace and deliver attacks. See Figure 10.10.

Penetrating Surviving Defenses. According to PLA writings, attacking aircraft would have the support of dedicated SEAD and EW aircraft. The mission of the SEAD aircraft would be to use passive electronic sensors to locate and destroy any SAMs and AAA that survived the initial missile barrage. The EW aircraft would attempt to jam surviving early warning and GCI radars, as well as radio and electronic data network communications—this sort of mission is often referred to in Western discussions of air operations as standoff jamming to distinguish it from the self-protection jamming systems most modern tactical aircraft carry, which are designed to degrade target-tracking and engagement systems. Figure 10.11 shows how these specialized aircraft might be positioned to help attacking aircraft break through the surviving "outer crust" of Taiwanese air defenses. (Note: the EW aircraft orbits are labeled "SOJ," for standoff jammer.)

Figure 10.11
Possible Location of Standoff Jammer and Suppression of Enemy Air Defense Orbits

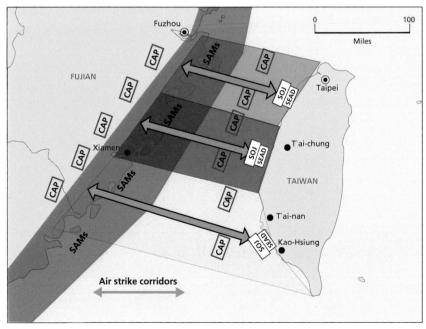

Strike Package Tactics and Configuration

In addition to being concentrated in space as a result of procedural deconfliction and control procedures, PLA air attacks would likely be concentrated in time as well as by scheduling the arrival of attacking aircraft so that they combined into strike packages. As noted in the section "Strike Methods and Force Composition" in Chapter Five, PLA publications refer to this as a "concentrated strike." This has the advantage of presenting a large number of targets to a limited number of defending systems over a very short period of time. The number of attacking aircraft any given array of defending systems can effectively engage and destroy is limited by the amount of ready munitions they carry. In most cases, the ammunition in question will take the form of missiles—either SAMs or air-to-air missiles. For example, a system with eight ready missiles and tactics that dictate firing two missiles at each target can engage four attackers before it must reload. If it takes 20 minutes to reload, but only 12 minutes for an attack formation, or strike package, of 40 to 50 aircraft to pass by, then it can engage only four attackers. Contrast this with a "smooth flow" or "bomber stream" attack pattern, referred to as a "continuous strike" in PLA publications, in which attackers arrive at a more or less steady rate over the course of an hour. In that case, there would be one aircraft flying past the defending system every 70 to 90 seconds. The defender could engage four aircraft in about five minutes, reload and then engage another four aircraft, then reload and engage four more, for a total of 12 engagements against the same number of bombers. So, in this example, concentrating the aircraft into a strike package denies the defender two-thirds of its possible engagement opportunities.

There are other advantages to a concentrated strike as well. Different sorts of aircraft, or similar aircraft carrying different types of specialized weapons or other systems, can be mixed and assigned specific mutually supporting missions that maximize the probability of overall mission success and make the strike package more than the sum of its parts. For example, specialized SEAD aircraft might lead the way to suppress any surviving enemy ground-based air defenses. Following quickly behind the SEAD aircraft might be one or more elements of air superiority fighters assigned to engage any enemy aircraft in the target area and establish an OCA CAP. If all went as planned, these two lead elements of the strike package would secure the airspace in the target area before the main body of the formation arrived. This would consist primarily of a large number of fighter-bombers or attack aircraft with a few air superiority fighters as close escort. Once the attack aircraft finished delivering their munitions, the formation would leave in reverse order, with the main body departing immediately, followed within a minute or two by the OCA CAP and finally the SEAD aircraft. Figure 10.12 shows the illustration in Chapter Five from the *China Air Force Encyclopedia* depicting a typical strike package and, for comparison, an illustration of a typical USAF low-level strike package from the 1980s.

It is worth noting that, in recent years, the PLAAF has begun to emphasize night formation flying at low altitude over the ocean.[22] This indicates the willingness to accept the likelihood of significantly increased operational training losses in order to begin building an experience base among PLAAF fighter and attack aircraft crews that will support the sort of low-level strike package operations described in this monograph. This is another indication that the PLA is serious about implementing the capabilities and concepts described in its publications.

Moving Forward

As ground-based air defenses on Taiwan were progressively destroyed and Taiwan's air force worn down by attacks on its bases and air-to-air combat attrition, it is likely that the PLA would move its OCA CAPs forward to positions over Taiwan itself. Analysis conducted for this study suggests that it could take at least a week before this move was

---

[22] See Allen, 2005b. This article states, in part,

> On September 26, 2004, PLA Daily carried an article with the title "Breakthroughs Made in Night Maritime Flight Training." The training focuses on "boosting the pilots' psychological quality and technical and tactical skills." The article also states, "Pilots conducted repeated exploration of fighting methods in combination with highly difficult flying training subjects, such as low altitude flying[,] and upgraded their training from former simple flight training to comprehensive training which integrates skills and tactics with fighting methods, making training much closer to real air battles."

Figure 10.12
Chinese Doctrinal Strike Package Compared to USAF Low-Level Strike
Package of the 1980s

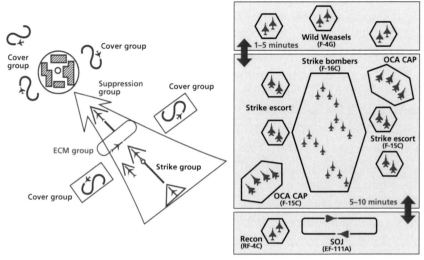

PLAAF doctrinal strike package   Typical 1980s USAF package

SOURCE: PLAAF, 2005.
RAND MG915-10.12

made.[23] From such a position, the PLA could further suppress Taiwan's air operations. It is likely that, at this point, supporting assets, such as AWACS and tankers, would also move forward to locations over the Taiwan Strait.

Su-27/J-11 aircraft operating over eastern Taiwan and armed with very long-range air-to-air missiles would pose a serious threat to USAF and U.S. Navy (USN) ISR aircraft and tankers. By the 2015–2020 time frame, it is possible that PLAAF Su-27/J-11 aircraft will be armed with one or more of the following Russian (or Chinese equivalent) very long-range air-to-air missile systems: the Kh-31, the R-77M, or the R-172. The Kh-31 is a ramjet-powered antiradiation missile currently

---

[23] This analysis assumed that attacks on Taiwan would be preceded or accompanied by air and missile attacks on U.S. air bases in the region.

in production and is a standard store on Russian Su-30s.[24] It can be equipped with either a passive antiradiation seeker or an active radar seeker with midcourse data-link updates. It has a maximum range of about 60 nm. At about 1,300 pounds, it is a large missile, but the big Su-27/J-11s can carry up to six.[25]

The R-77M is a ramjet-powered derivative of the R-77 (AA-12 "Adder") missile. The use of ramjet propulsion increases range from about 54 nm for the standard rocket-propelled R-77 to about 86 nm for the R-77M. Weight is about 500 pounds, compared to less than 400 pounds for the standard R-77. This missile is currently in production in Russia and uses active radar homing with midcourse data-link updates.

Finally, there is the possibility that at least some of the OCA CAP aircraft might carry the 1,650-pound R-172. This is a new Russian missile and may not yet have reached production status.[26] However, it is quite possible that, over the next 10 to 15 years, something like it will become available for the PLAAF's Su-27/J-11 fleet. This weapon is designed for use against large, unmaneuverable targets, such as U.S. ISR platforms. It has a claimed range of 215 nm. Even if its effective range is only 80 percent of that claimed, it will be a formidable anti-ISR weapon.[27]

Operating over eastern Taiwan, OCA CAPs armed with weapons like these could, without even leaving their OCA CAP locations, force U.S. airborne ISR platforms to remain at least 150–300 nm (or more) away from the Taiwan Strait. Tactics that included occasional sweeps

---

[24] The Kh-31 is primarily a supersonic antiship missile but could be modified for the air-to-air antiradiation missile role. Since China already possesses Su-30s and antiship versions of the Kh-31, China could integrate an air-to-air version of the Kh-31 into its fighter fleet in a reasonably short time if it desired to field this sort of capability.

[25] It is more likely that aircraft flying OCA CAPs over Taiwan would carry only two of these large missiles along with a mix of R-77 medium-range and R-73 short-range air-to-air missiles. The same applies to the R-172.

[26] In 2007, it was being actively marketed by the Russians on the Su-35BM and Su-35-1; see "Air Power Australia," 2007.

[27] See "Air Power Australia," 2009, for a more detailed description of Su-27/J-11 derivative capabilities, growth paths, and weapon options.

forward from the OCA CAP locations could force ISR assets even further back.

While a Global Hawk (RQ-4) operating at 60,000 feet would still theoretically have radar line of sight to surface-ship movements in the Taiwan Strait while operating 150 nm from the west coast of Taiwan, at 250 nm or more, it would not. AWACS (E-3) and JSTARS (E-8) operating between 30,000 and 40,000 feet would be unable to see low-level air or surface traffic in the strait if forced to operate beyond about 175–210 nm from the west coast of Taiwan.

Figure 10.13 shows the possible location of PLA OCA CAPs over Taiwan and the ISR exclusion zone that such CAPs might create using missiles like those the PLAAF already possesses, such as the standard R-77. Figures 10.14 and 10.15 show how this might increase with the acquisition of weapons like R-77M or the R-172.

Figure 10.13
Area Where PLAAF OCA CAPs Operating over Eastern Taiwan Could
Effectively Threaten USAF ISR Assets Using Standard R-77 Missiles

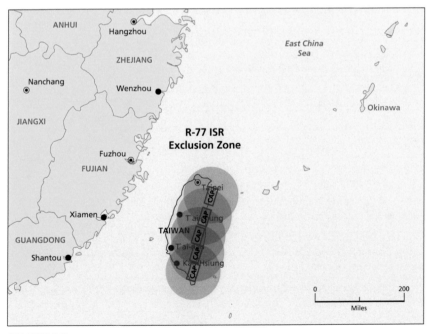

Figure 10.14
Area Where PLAAF OCA CAPs Operating over Eastern Taiwan Could
Effectively Threaten USAF ISR Assets Using Currently Available R-77M
Missiles

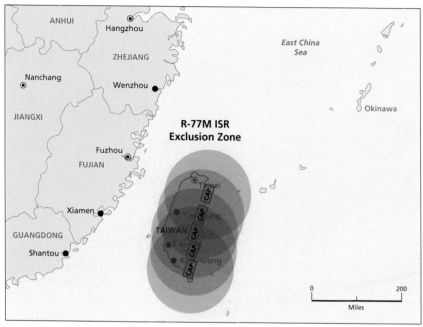

If PLA fighters can effectively deny U.S. airborne ISR assets the ability to detect and track air and sea movement across the Taiwan Strait, they will have laid a foundation for an invasion of Taiwan by the PLA. In addition, air-to-air refueling operations could be severely disrupted due to the need to remain well outside the very long-range air-to-air missile footprint. This could severely constrain both USAF and USN air operations and might be compounded by the limited number of airfields not under threat of attack by land-attack cruise missiles or TBMs. Due to the geography of the western Pacific, the United States does not have the strategic depth or number of basing options it enjoyed during the Cold War in Europe or more recently in the Persian Gulf. Therefore, there is a risk that the few available tanker

Figure 10.15
Area Where PLAAF OCA CAPs Operating over Eastern Taiwan Could Effectively Threaten USAF ISR Assets Using R-172 Missiles Currently Under Development

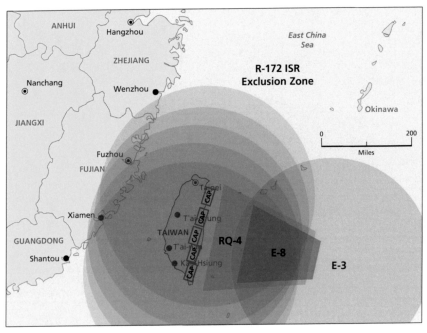

RAND *MG915-10.15*

bases would be overtaxed in fuel-replenishment capability or run out of tarmac area.

If PLA operations reached this stage before the United States could bring sufficient forces into theater to effectively hinder PLA operations, or if operations from Guam and Japanese bases were disrupted by effective preemptive strikes, the probability of a successful invasion of Taiwan would be greatly increased.

Observations on China's Air Offensive Capabilities and Concepts

Overall, Chinese strategy, along with its writings on air offensive campaigns, stresses the importance of defending key Chinese assets while seizing the initiative early in a conflict through the use of surprise and preemption. Over the past 15 years, the PLAAF has made great strides in moving from a huge, largely defensive force of mostly obsolete aircraft to a smaller (though still quite large) force of much more modern aircraft.

Trends in training, equipment acquisition, and tactics indicate that, over the next ten years, the PLA is likely to acquire a large number of fourth-generation fighters with increasingly well-trained crews. In addition, it seems to be investing in significant new theater strike capabilities centered on a large force of H-6–derivative cruise missile–carrying aircraft. These could be supplemented or escorted by Su-27/J-11 multirole fighters as the PLA acquires air-to-air refueling aircraft. There is every indication that the PLA is intent on transforming itself into a force capable of carrying out all of its doctrinal air force campaigns. In addition, recent trends in the acquisition and license-production of advanced Russian and other foreign weapon and electronic systems, and especially the development of the H-6K, indicate that PLA planners are as creative and resourceful as they are determined.

Overall, the PLA is rapidly developing a modern, capable air force. While there will undoubtedly be false starts and hard-won lessons over the next 10 to 15 years as the PLA develops and digests a number of new operational concepts, tactics, and systems, the experience of the past decade or two indicates that it is likely to be quite successful in bringing new capabilities on line. If this is the case, the PLA will begin to present formidable challenges to U.S. air operations in the western Pacific, especially over and around Taiwan, by the latter part of the next decade.

Implications for the U.S. Air Force

This chapter has described how the emerging technical capabilities and operational competence of the PLA could pose a number of challenges to U.S. air operations over the next decade or two. Prominent among these is the "antiaccess" threat potentially posed by a fleet of H-6 cruise missile carriers to distant-theater bases that have until now been assumed to be beyond the effective reach of Chinese conventional forces.

If the PLAAF begins to deploy a large force of land-attack cruise missile–carrying H-6 bombers, the USAF would need to be prepared to implement a set of cruise missile defensive measures for Andersen and other key facilities in the western Pacific that would be capable of defeating, or at least greatly neutralizing, a large-scale cruise missile attack, even if it is launched as a preemptive surprise attack. Such a system would ideally be capable of shooting down the missile-carrying aircraft before they launched their missiles.

In addition to the threat of cruise missile attacks on Guam, moreover, is the possibility that China will deploy conventional ballistic missiles capable of reaching the island. China clearly has the technical capability to develop such a missile, as it already has in service the road-mobile, solid-fuel DF-31 and DF-31A missiles. These missiles have ranges (roughly 4,000 nm and 6,000 nm, respectively) significantly greater than would be needed to strike Guam, which is about 1,600 nm from mainland China, although they are currently dedicated to carrying nuclear payloads.

The potential vulnerability of bases as far as 3,000 nm from mainland China is compounded by the unique geography of the western Pacific. Because islands in this region are generally few and far between, it is often not possible to fall back gracefully from a given threatened base to a new one just outside threat range.

Other implications for the USAF of growing PLA capability and competence in the context of a conflict over Taiwan include the likelihood that the magnitude of the air threat to Taiwan would be such that few USAF fighter sorties could be used for any missions besides DCA and SEAD for the first weeks of the conflict. This would limit USAF offensive capacity against Chinese facilities to a handful of B-2 sorties per night. Finally, in addition to taking a significant amount of time to defeat, PLA operations against Taiwan in the 2015–2020 time frame stand a good chance of inflicting severe damage on Taiwanese military forces and civilian infrastructure and inflicting significant losses on USAF and USN aircraft opposing them.

CHAPTER ELEVEN

Conclusions and Implications

The most immediate observation that suggests itself from the analysis of Chinese military publications on air force operations as described in the preceding chapters is how systematic and comprehensive they are. Few militaries in the world have such extensive published documentation on the employment of air forces. The concepts described, moreover, appear to be realistic and practical, drawing on the experience of other air forces in recent conflicts, particularly those of the United States (the PLAAF having had no significant combat experience since the 1950s), but remaining appropriate to the current and near-future capabilities of the PLAAF. Chinese military analysts are clearly engaged in a serious process of developing specific, practical concepts for the employment of China's air forces. This in itself is a significant finding: If this concept development is reflected in actual training and, in the event of a conflict, in campaign and mission planning, the United States would find itself engaged with adversary air forces both qualitatively and quantitatively superior to those it has fought since the end of the Cold War. Indeed, the United States has not fought a conflict against an adversary capable of challenging its supremacy in the air since at least the Korean War.

A second observation is that, although the PLAAF has traditionally emphasized defensive operations, that is no longer the case, and the United States would likely find the PLAAF to be an aggressive opponent in the event of a conflict. The PLA clearly prefers to achieve air superiority by attacking its enemy on the ground or water. Especially at the beginning of a war, the PLA will endeavor to attack enemy air bases, ballistic missile bases, aircraft carriers, and warships equipped with land-attack cruise missiles before enemy aircraft can take off or enemy missiles can be launched. Thus, the PLAAF can be expected to carry the fight to the United States in the form of direct attacks on U.S. air bases and ships. These attacks, moreover, will be carried out not by China's air force operating in isolation but in coordination with the Second Artillery's conventional ballistic and cruise missiles. As a consequence, for the first time since the end of the Cold War, U.S. air forces would not be able to regard their bases as sanctuaries safe from enemy attack in a conflict. This threat of attack should not be understood as simply a missile threat. Rather, it would be a joint aerospace threat in which ballistic missiles would be a critical enabler for more-precise land-attack cruise missiles and PGMs carried by manned aircraft. This means that U.S. forces must once again plan seriously for the defense of air bases against attack by not only ballistic missiles but also aerodynamic threats.

Offensive operations against China would be challenging as well, as Chinese military publications emphasize defensive operations even in an offensive air campaign. The PLA's concept of layered air defense, when combined with China's strategic depth, its highly capable fighter interceptors and mobile SAMs, and its emphasis on hardening, camouflage, and concealment, would make strike operations over Chinese territory high-risk propositions for nonstealthy aircraft. Hardened shelters and the large number of military airfields in China, moreover, mean that China's air forces cannot easily be destroyed on the ground as were Egypt's in 1967 or Iraq's in 1991. To be effective, strikes against targets in mainland China would require (1) significant effort devoted to suppressing China's long-range SAMs, (2) stealthy delivery platforms and long-range standoff munitions, and (3) large numbers of sorties, and should be expected to incur significant losses on the U.S. side.[1]

---

[1] The overlapping coverage and frequency diversity of the Chinese early warning system, coupled with modern SAMs and fighters employed in large numbers, will make for a very challenging environment even for stealthy aircraft. Although low-altitude operations may offer some potential for evading these threats, the risk level and potential need to strike deep targets without the possibility of fighter escort mean that the long-range standoff munition option is likely to be preferred. Moreover, ideally these munitions will themselves be stealthy.

Beyond these general observations, Chinese concepts for the employment of air forces, as described in Chapters Four through Nine, combined with the air, missile, and munition capabilities China is acquiring, have more-specific implications when applied to particular contingencies. Since by far the most likely conflict involving the United States and China would be one over Taiwan, it is useful to review how China's air force employment concepts and capabilities might be implemented in such a conflict.

China's Air Force Employment Concepts and Capabilities in a Conflict Over Taiwan

If the PRC chose to use force against Taiwan, whether in the form of an outright invasion or a blockade, it would likely begin with an offensive air campaign against the island. As described in Chapter Five, the preparatory step for such a campaign would be information reconnaissance. This would include mapping Taiwan's civilian and military information systems using publicly available information, information acquired through covert network intrusions, and information acquired through traditional forms of espionage. It would also include collecting information on the location, frequencies, and modes of Taiwan's early warning, command-and-control, SAM, and other sensors and communication systems. Some of the information would be collected through human espionage and some using electronic intelligence (ELINT) aircraft.[2]

The actual campaign would start with an information offensive. This would entail computer network attacks, electronic deception, electronic interference, and firepower destruction. Prior to the launching of physical attacks on Taiwan, computer network attacks would likely take the form of covert efforts to disable Taiwan's early warning systems and communication networks and to insert software exploits

---

[2]   Such as a possible ELINT version of the Y-8 transport aircraft or the ELINT UAV advertised by the China Aerospace Science and Industry Corporation at the 2006 Zhuhai air show. See "SAC Y-8/Y-9," 2009; "Chinese Signals Intelligence," 2007.

for later use. Once physical attacks began, computer network attacks would probably include more-aggressive efforts to penetrate and exploit or disable all of Taiwan's military information and communication systems through the insertion and activation of backdoors and viruses and through denial-of-service attacks.

Electronic deception prior to the launching of physical attacks would likely take the form of suppressing or disguising as civilian activities any electromagnetic emissions that might indicate that an attack was imminent. Once the physical attacks were under way, electronic deception would presumably take the form of false or misleading emissions suggesting that forces were present in locations where they actually were not, or suggesting that they were of a type other than they actually were, and suppressing the emissions of actual forces. Active electronic interference would probably not be initiated until just prior to the launching of physical attacks and would be directed against reconnaissance and early warning satellites (in the form of laser dazzling in the case of electro-optical satellites, and possibly including U.S. and commercial imagery satellites used by U.S. or Taiwanese military or intelligence agencies, even if destructive attacks against U.S. forces were not planned initially), airborne early warning and control aircraft, ground-based early warning radars, the radars of SAMs and interceptor aircraft, and radio communications.

Firepower destruction would consist of ballistic missile and cruise missile attacks on Taiwan's early warning and fire-control radars and other electromagnetic emitters, such as radio communication facilities.

Once the information offensive was launched, the penetration of enemy air defenses would be initiated. As described in Chapter Nine, this would include attacks by conventional ballistic missiles of the Second Artillery Force against Taiwan's air force bases, SAM batteries, and command-and-control facilities (in addition to the just-mentioned attacks on Taiwan's electromagnetic emitters as part of the firepower destruction component of the information offensive). As of late 2009, China had roughly 1,100 DF-11 (CSS-7) and DF-15 (CSS-6) conventional ballistic missiles capable of reaching Taiwan.[3] Even if only half

---

[3]   Office of the Secretary of Defense, 2010, p. 66.

of these missiles were targeted against Taiwan, they would easily overwhelm Taiwan's Patriot and Tien Kung missile defense systems. Not only would their sheer numbers exceed the number of missile interceptors Taiwan has; China is estimated to have at least 120 DF-11 launchers. It would be a relatively simple matter to time the firing of missiles from these launchers so that more than 100 missiles would arrive over Taiwan simultaneously. Taiwan is expected to have at most nine PAC-2 and PAC-3 Patriot batteries and 18 Tien Kung 2 and Tien Kung 3 batteries in the coming decade and thus could engage only a fraction of the incoming missiles, allowing the remainder to reach their targets unhindered.[4]

If the missiles launched at Taiwan possessed the types of warheads described in Chapter Nine as already being under testing, then warheads with fléchette submunitions would likely be used against ground-based missile-launch equipment (e.g., Patriot launchers), aircraft parked in the open, above-ground fuel tanks, and other "soft" targets. Warheads with penetrating submunitions would be used against airfield runways, aircraft hangars, and semiunderground fuel tanks; and unitary warheads or warheads with blast submunitions would be used against command centers, Tien Kung batteries (which are deployed in underground cells), and other fortified targets. The net result of these initial ballistic missile attacks, including the attacks on radars, radio communication facilities, and other electromagnetic emitters that would be part of the information offensive, would likely be that most or all of Taiwan's non-mobile SAM systems (i.e., Tien Kung and, if not frequently relocated, Patriot) would be rendered combat ineffective, many or most aircraft parked in the open would be damaged, the runways at Taiwan's roughly a dozen airfields capable of operating combat aircraft would be at least temporarily unusable, and many of

---

[4]   Office of the Secretary of Defense, 2010, p. 66; "DF-11 (CSS-7/M-11)," *Jane's Strategic Weapon Systems*, June 18, 2007; "Tien Kung 1/2/3 (Sky Bow)," *Jane's Strategic Weapon Systems*, February 22, 2008.

Taiwan's command, control, and communication facilities and early warning systems would be damaged or destroyed.[5]

The ballistic missile attacks would be followed by manned aircraft and cruise missile attacks. The first goal of these aircraft and cruise missiles would be electronic interference and suppression. SOJs (such as the apparent EW versions of the Y-8 airframe that have been seen) would attempt to make detection of the strike group more difficult by increasing the overall level of background noise in Taiwan's radar receivers, and JH-7 escort jammers would attempt to prevent the strike aircraft from being engaged by Taiwan's SAMs or by radar-guided air-to-air missiles launched from any of Taiwan's fighter aircraft that were able to get aloft. Russian-made Kh-31P and Chinese-made YJ-91 supersonic antiradiation missiles carried by Su-30, JH-7, and, in the future, multirole versions of the J-11 (based on the Su-27), as well as ground-launched antiradiation cruise missiles, such as the Israeli-made Harpy or an antiradiation version of the DH-10 ground-launched cruise missile, would engage any still-operational early warning, SAM, or other radars and other militarily significant electromagnetic emitters.[6]

In addition to suppression of Taiwan's electronic systems, China's aircraft and cruise missiles would attempt to suppress Taiwan's air defense firepower, which would involve attacking SAM and other air defense installations that had survived the earlier ballistic missile attacks, as well as attacking fighter bases and other key targets, such as command-and-control facilities. In addition to China's ground-launched DH-10 cruise missile, of which it is already estimated to have 200–500, the H-6 bomber can carry the YJ-63 ALCM (and probably an air-launched version of the DH-10 in the future), and several

---

[5] "Tien Kung 1/2/3," 2008; "China: Air Force," *Jane's Sentinel Security Assessment: China and Northeast Asia*, February 4, 2008.

[6] "Chinese Electronic Warfare (EW) Aircraft," *Jane's Electronic Mission Aircraft*, August 30, 2008; "SAC Y-8/Y-9," 2009; David A. Fulghum and Douglas Barrie, "Non-War: China Accelerates Focus on Disruption, Asymmetric Tactics," *Aviation Week and Space Technology*, March 10, 2008; "Sukhoi Su-30," 2008; "Sukhoi Su-27 Aircraft Corporation," 2008; "Xian Aircraft Company: XAC JH-7," *Jane's All the World's Aircraft*, August 2, 2007; "YJ-91," 2007; "IAI Harpy and Cutlass," *Jane's Unmanned Aerial Vehicles and Targets*, March 25, 2008.

types of Chinese aircraft are currently known to carry laser-guided and satellite-guided bombs, including the Q-5 attack aircraft and the J-8, JH-7, and Su-30 multirole aircraft, and, in the future, the J-11 and J-10 light fighter likely will as well. The PLAAF's current inventory reportedly includes up to 80 cruise missile–capable H-6s, 120 Q-5s, 50 J-8IIIs (the newest version of the J-8 air superiority fighter), 70 JH-7s, and 70 Su-30s. Even if only a fraction of the PLAAF's Q-5s are currently capable of carrying laser-guided or satellite-guided munitions, therefore, the PLAAF already has more than 200 aircraft capable of carrying PGMs, and this number will only increase in the future as China continues to produce H-6s, JH-7s, and J-8IIIs and begins producing multirole versions of the J-11 and J-10. These aircraft and missiles would attack any known SAM sites, as well as shelters for fighter aircraft (unsheltered aircraft presumably having been largely destroyed by the initial ballistic missile attacks), command-and-control facilities, aviation fuel storage and distribution facilities, and repair and maintenance facilities at Taiwan's fighter bases.[7]

While Taiwan's air defense firepower was suppressed, air strikes against the primary targets of the offensive air campaign then would be carried out. The first objective of an offensive air campaign being to seize air superiority, however, initially those primary targets would in fact also be Taiwan's air defense capabilities, particularly its early warning, command-and-control, and communication facilities; its SAMs; and its fighter aircraft and bases as described earlier; but other targets of the air offensive campaign would include any other military aircraft and air bases, SSM batteries, and any other targets associated with Taiwan's ability to conduct air or missile operations. Once those forces and facilities were largely destroyed, air and ballistic and cruise missile attacks would shift to other targets (while continuing at a lower level against Taiwan's air and missile targets, to ensure that Taiwan did not recover its capability to conduct operations with those forces). As described in Chapter Five, these would include attacks on Taiwan's

---

[7] Office of the Secretary of Defense, 2010, p. 66; "KD-63," 2008; "LT-2 Laser Guided Bomb," 2007; "Fei Teng Guided Bombs," 2010; IISS, 2008, p. 48; Cliff, Burles, et al., 2007, pp. 51–64.

government, water, and electric installations, and, if a prolonged campaign were expected, then attacks would be made on other economic targets in Taiwan as well. If the offensive air campaign was part of an air blockade campaign as described in Chapter Seven, attacks would also shift to Taiwan's civilian and military transport aircraft and facilities. If the offensive air campaign was part of a joint blockade campaign, these attacks would include Taiwan's military and civilian sea ports and associated facilities, as well as naval and merchant ships at sea. (Any naval forces in port would likely be attacked as part of the initial missile and air strikes, but the PLAN aviation forces presumably would have primary responsibility for attacking ships at sea, as the PLAAF has limited capabilities for this mission.) If the offensive air campaign was part of a joint landing campaign, on the other hand, after air superiority was achieved, attacks would shift to Taiwan's naval bases, naval forces at sea (again, any naval forces in port having likely been attacked as part of the initial strikes), coastal defenses, and ground forces.

When conducting air strikes, both manned aircraft and cruise missiles would likely fly to Taiwan at low altitudes to evade detection by any still-operational early warning and engagement radars (though this would not prevent them from being detected by any of Taiwan's six current E-2 airborne early warning aircraft that survived the initial ballistic missile attack and managed to get aloft, or were already aloft at the time of the ballistic missile attacks). The attacking aircraft and cruise missiles would most likely approach Taiwan from directions other than the most direct route from their launch points, seeking instead to find gaps in Taiwan's radar coverage and air defenses on its northern, southern, or eastern sides. The strike groups would attempt to avoid engagement with any fighter aircraft that intercepted them, which would instead be engaged by the cover groups consisting of Su-27, J-10, J-8, and J-7 (MiG-21) aircraft. Even the cover groups, moreover, would attempt to avoid an extended engagement with intercepting fighters so as to avoid depriving the strike group of air cover.

If the air offensive campaign was part of an air blockade or joint blockade against Taiwan, once air superiority was achieved, China would also then establish air surveillance zones around the island as described in Chapter Seven. Sustaining surveillance zones on the eastern side of the island—the side from which aircraft running the block-ade would most likely attempt to ingress and egress—would require substantial loiter capabilities on the part of the surveilling aircraft or else their frequent replacement by newly sortied aircraft. China's long-range Su-30 and Su-27 aircraft would appear to be particularly well suited for this mission, but aerial refueling could be used to extend the loiter times of shorter-range aircraft, such as the J-10. In any case, given Taiwan's geographic isolation, once Taiwan's air force was defeated, absent U.S. intervention, relatively few fighters would be needed to enforce the air blockade.

If the air offensive campaign was part of a landing campaign against Taiwan, it is possible that an airborne campaign might be con-ducted as well. Given the requirement, stated in Chinese military pub-lications on airborne campaigns, to seize and maintain air superiority between the air bases from which the airborne troops would depart and the landing zone, and the requirement that the landing zone be lightly defended, an airborne operation would likely occur after China had largely defeated Taiwan's air forces or in a location well away from any surviving Taiwan air forces, and would not occur in a location with significant ground defenses. This could mean an airborne landing on one of Taiwan's outlying islands (such as the Penghu Islands) or in southern or eastern Taiwan, possibly in concert with an amphibious landing on a nearby beach to draw away defending forces.

If the United States chose to intervene in a Chinese attack on Taiwan, China would likely conduct a similar offensive air campaign against U.S. forces in the western Pacific.[8] As it would against Taiwan, such a campaign would be preceded by information reconnaissance efforts against U.S. civilian and military information systems and U.S. early warning, command-and-control, SAM, and other sensors and communication systems. The campaign itself would begin with an information offensive entailing computer network attack, electronic deception, and electronic interference. This would be followed by fire-

---

[8] Indeed, if U.S. intervention were viewed as inevitable, it is possible that China would launch an air offensive campaign against U.S. forces prior to or simultaneously with its attacks on Taiwan. See Cliff, Burles, et al., 2007, pp. 29–34.

power destruction of U.S. early warning and fire-control radars and other electromagnetic and information targets, such as radio communication facilities and terrestrial communication cables and switching facilities in the western Pacific.

Once the information offensive was under way, the penetration of U.S. air defenses would then begin. As in the case of Taiwan, destructive attacks in the information offensive and penetration of U.S. air defenses would be conducted both by the PLAAF's manned aircraft and by conventional ballistic and ground-launched cruise missiles of the Second Artillery. The newer versions of the DF-15 ballistic missile are believed to be capable of reaching Okinawa, as are conventional versions of the 2,000 km–range DF-21 mobile missile. Although China apparently does not currently have a conventional ballistic missile capable of reaching Guam (3,000 km from mainland China), since it has already fielded road-mobile intercontinental ballistic missiles (ICBMs) with ranges of more than 7,000 km, China clearly has the technical capability to develop conventional mobile ballistic missiles capable of doing so.[9] Even if only one-half of the 350–400 DF-15 missiles China possesses are capable of reaching Okinawa, China has at least 175 missiles that could be used to attack the three military air bases on the island (Kadena Air Force Base, Marine Corps Air Station Futenma or its replacement, and the Japan Air Self Defense Force base at Naha). And even if China has only 50 launchers capable of firing the longer-range version of the DF-15 (China is estimated to have a total of approximately 100 DF-15 launchers, but it is unclear how many are capable of launching the longer-range version), it would be a relatively simple matter to time the launchings so that at least 40 warheads arrived over Okinawa nearly simultaneously, easily overwhelming the three PAC-3 batteries currently on the island. As in the case of Taiwan, therefore, the likely result of the initial ballistic missile attack on Okinawa would be that many or most aircraft parked in the open would be damaged, the runways at the attacked airfields would be at least temporarily unusable, and many command-and-control facilities and early

---

[9] There are reports that China is developing a mobile ballistic missile with a range of 2,500–3,000 km. See "DF-25," *Jane's Strategic Weapon Systems*, June 29, 2007.

warning systems on Okinawa, as well as any SAM units whose position was known prior to the attack, would be damaged or destroyed.[10]

The ballistic missile attacks would be followed by manned aircraft and cruise missile attacks. China's ground-launched DH-10 cruise missiles and H-6, JH-7, Su-30, Su-27, and J-11 aircraft without aerial refueling are all capable of reaching Okinawa. As in the case of Taiwan, aircraft and cruise missiles would likely fly at low altitudes to evade detection by any still-operational land-based early warning and engagement radars on Okinawa (although they would, of course, be subject to detection by any airborne radars that were aloft at the time of the initial attack) and would approach the island from directions other than the most direct route from their launch points. Standoff Y-8 jammers would attempt to make detection of the strike group more difficult by increasing the overall level of background noise in Okinawa's radar receivers, while JH-7 escort jammers would attempt to prevent the strike aircraft from being successfully engaged by any surviving SAMs or by radar-guided air-to-air missiles launched from any U.S. or Japanese fighter aircraft that had managed to get or stay aloft. Air-launched Kh-31 and YJ-91 antiradiation missiles and possibly ground-launched antiradiation cruise missiles would engage any still-operational early warning, SAM, and other radars and other militarily significant electromagnetic emitters. As noted, China currently has 80 H-6, 70 JH-7, and 70 Su-30 aircraft and will soon begin producing a multirole version of the J-11, all of which could then use a variety of PGMs to attack any known SAM sites, as well as aircraft shelters, command-and-control facilities, aviation fuel storage and distribution facilities, and repair and maintenance facilities at Okinawa's air bases. Since the goal of attacking military facilities on Okinawa would simply be to prevent U.S. (and Japanese) forces there from intervening in a Chinese use of force against Taiwan, rather than to impose a blockade on or invade Okinawa or the rest of Japan, attacks on Okinawa would probably be limited to targets associated with its ability to contest air and sea superiority around Taiwan. Similar attacks might also be conducted against

---

[10] "DF-15 (CSS-6/M-9)," *Jane's Strategic Weapon Systems*, July 6, 2007; "DF-21," 2007; Office of the Secretary of Defense, 2008, p. 56.

U.S. military facilities (including naval facilities and ships in port) on the main islands of Japan or elsewhere in the western Pacific, such as Guam, although the number of Chinese aircraft and missiles able to reach some of these targets would be significantly smaller.[11]

In addition to the threat posed to U.S. and Taiwanese air forces in the western Pacific by Chinese air attacks, offensive air operations *against* China would also be challenging in a Taiwan scenario. In a Chinese offensive air campaign, a significant portion of China's aircraft would be reserved for resistance operations, and China has increasingly formidable SAM capabilities.[12] According to the Chinese military publications analyzed in Chapters Five and Six, resistance operations would entail China keeping a portion of its fighters in the air at the most forward point practical in the expected direction of an enemy air attack. The role of these aircraft would be to engage and delay incoming strike packages, allowing additional interceptors still on the ground to scramble into the air. Once the incoming U.S. strike package made it past these initial interceptors, moreover, it would be engaged by long-range SAMs deployed along China's coast. As of late 2008, the PLAAF had 24 batteries (each with eight quadruple launchers) of modern, long-range SAMs with maximum intercept ranges of 100–200 km and was continuing to acquire more. Even if only half of them were deployed in support of an offensive air campaign against Taiwan (as noted in Chapter Six, at least a portion of them would likely be deployed to protect the capital, Beijing), they could form a dense belt of overlapping fields of fire along China's southeastern coast. Any incoming strike aircraft or cruise missiles that managed to penetrate this belt would then be subject to intercept by the fighters that had been scrambled during the initial engagement. Finally, before reaching their targets, any surviving U.S. aircraft, missiles, or munitions

---

[11] "Xian Aircraft Industries Group," 2007; "Xian Aircraft Company," 2007; "Sukhoi Su-30," 2008; "Sukhoi Su-27," 2008.

[12] As noted in Chapter Five, at most, 80 percent of *strike* aircraft would be committed to even the initial strikes, so presumably an even lower percentage of air-to-air interceptors would be committed during the initial strikes, and the number of interceptors reserved for resistance operations would undoubtedly further increase after the initial strikes had been carried out.

would be subject to intercept by gun and SAM point defenses, including the highly capable Tor system (SA-15), as well as to efforts to present false targets and to conceal, camouflage, and fortify real targets, as illustrated in Chapter Ten. In addition, according to the publications analyzed in Chapter Five, Chinese combat aircraft will likely operate from a multiplicity of airfields, including not just the air bases they use in peacetime but also civilian airports, reserve airfields, and decommissioned airfields. Aircraft will be deployed in depth, with fighter aircraft that have offensive roles operating from the airfields closest to the coast, attack aircraft and fighter-bombers deployed at airfields farther inland, and bombers, airborne warning and control aircraft, aerial refueling aircraft, and the campaign reserve deployed deep inland. In an OCA campaign, therefore, U.S. strikes would have to reach deep into China to disable its attack, fighter-bomber, and bomber forces, even in a nominally localized war, such as a conflict over Taiwan.[13]

Implications for the United States

The above analysis has a number of implications for the United States. First, if the United States intervenes in a conflict between the PRC and Taiwan, it should expect attacks on its forces and facilities in the western Pacific, including those in Japan. These attacks would likely not be restricted to ships at sea and aircraft in the air. Chinese military publications on the use of airpower indicate a clear preference for attacking an enemy's air forces on the ground, and, if U.S. forces based in Japan were engaged in combat with Chinese forces, then both the United States and Japan would be regarded as belligerents under international law. Thus, it would be imprudent to assume that U.S. bases in Japan or elsewhere would be sanctuaries from Chinese attack in the event of a Chinese use of force against Taiwan and to deploy forces based on that assumption of inviolability. The effectiveness of such attacks,

---

[13] Office of the Secretary of Defense, 2009, p. 66; "SA-10/20 'Grumble' (S-300, S-300 PMU, Buk/Favorit/5V55/48N6," *Jane's Strategic Weapon Systems*, December 29, 2006; "HQ-9/-15," 2008; "Tor," *Jane's Land-Based Air Defence*, April 13, 2010.

moreover, would be enhanced if the United States were not expecting them and failed to park aircraft inside shelters, to continuously keep early warning and interceptor aircraft airborne, or to regularly relocate SAM batteries. Chinese military writings, moreover, emphasize the advantages of preemptive and surprise attacks, so it is possible that Chinese attacks on U.S. forces in the western Pacific would precede a use of force against Taiwan.[14]

Even in the absence of a crisis over Taiwan, therefore, the United States should also take steps to prevent China from collecting information on military and sensitive civilian information systems or on U.S. early warning, command-and-control, SAM, and other sensors and communication systems. At the same time, U.S. intelligence collectors should expect extensive efforts to deceive them about the locations and posture of Chinese forces. Indeed, any evidence that heightened deception efforts were under way would be an indication that an attack was being prepared. In addition, U.S. forces should ensure, to the maximum extent practical, that their information systems are protected from network intrusions, some of which may be going on today, or denial-of-service attacks.[15] It should also plan and train for the possibility that some of these systems will fail or be compromised in a conflict. In addition, the United States should be prepared to deal with electronic jamming on a scale larger than it has seen in any conflict since the end of the Cold War.

The United States should accept the likelihood that the runways of Okinawa's military airfields will be rendered at least temporarily unusable and that many or most unsheltered aircraft will be damaged or destroyed in the initial salvo of ballistic missiles. Sheltered aircraft, fuel storage and distribution facilities, and repair and maintenance facilities will then be vulnerable to follow-on attacks by cruise missiles and manned aircraft with PGMs. One set of responses to this challenge would be to increase the number of missile defense systems from the current three PAC-3 batteries, in the hopes of at least thinning out the incoming missiles and increasing the likelihood that at least some

---

[14] See Cliff, Burles, et al., 2007, pp. 29–34.

[15] Office of the Secretary of Defense, 2008, pp. 3–4.

runways will remain usable; to build shelters capable of protecting all aircraft to be based on Okinawa; to harden runways and fuel and repair facilities; and to increase rapid runway repair capabilities. Mobile point-defense systems, such as the U.S. Army's Surface-Launched Advanced Medium-Range Air-to-Air Missile (SLAMRAAM), could help defend Okinawa's air bases against aircraft, cruise missiles, and PGMs, increasing the number of sheltered aircraft that would survive the cruise missile and aircraft attacks that would follow the initial ballistic missile salvo while runways were not yet repaired. And, given the possibility that an attack could come with little warning, if even vague indications are received that China might be planning to use force somewhere in East Asia, the United States should begin parking aircraft in shelters when not in use, begin keeping early warning and interceptor aircraft continuously airborne, and regularly relocate its SAM batteries to unpredictable sites.[16]

An alternative approach would be to keep relatively few combat aircraft on Okinawa in the event of a crisis over Taiwan and instead deploy the bulk of U.S. land-based air forces to several more-distant bases in Japan and elsewhere in the western Pacific. Given China's capability to build mobile ballistic missiles of longer range than the DF-15, even more-distant bases should not be regarded as sanctuaries. China will be able to build fewer such missiles, however, particularly in the near term, and, if U.S. combat aircraft operate out of multiple airfields, the number of missiles that China can fire at each might be reduced to the point at which active missile defenses, aircraft shelters, hardened runways and facilities, and rapid runway repair capabilities would be sufficient to keep at least some of them viable.

The main islands of Japan are more than 800 nm from Taiwan, however, and Guam is nearly 1,500 nm away. If U.S. land-based aircraft are forced to operate out of bases that are so distant, it will be difficult to continuously maintain large numbers of fighter aircraft in the air over Taiwan.[17] The short distances between mainland China

---

[16] "HUMRAAM/SLAMRAAM," *Jane's Land-Based Air Defence*, October 22, 2008.

[17] This study did not assess threats to aircraft carriers, but it is possible that the combination of submarines, land-based aircraft, supersonic antiship cruise missiles, and antiship ballistic

and Taiwan, conversely, mean that China could amass a large number of aircraft on short notice. Thus, U.S. fighters would likely face overwhelming odds in engagements to defend Taiwan's airspace. One implication of this is that any fighter aircraft used to defend Taiwan must be capable of defeating several times their number of Chinese fighters. The performance capabilities of the F-22, coupled with the superiority of U.S. pilots and command and control, may provide such an advantage today, but the USAF will need to continue to invest in technology and pilot skill to ensure that it maintains its advantage in the face of rapid Chinese improvements in these areas.

It may also be that tactical fighter aircraft are not the optimal platform for providing air defense in locations so far from the nearest viable air base. An alternative, or supplement, might be a larger aircraft capable of carrying a large number (e.g., 20 or more) of extremely long-range (e.g., 200 nm) air-to-air missiles. Such an aircraft could engage Chinese fighters while still beyond the range of their missiles and then withdraw before it could be engaged by any of the survivors. An aircraft such as the B-1, which has a payload of 75,000 lb and supersonic dash capability, would be one possibility for providing this capability. A stealthy aircraft like those that were considered for the USAF's now-canceled Next Generation Bomber program would be another, particularly if the air-to-air missiles it carried had active seekers, and a more survivable aircraft, such as the F-22, could provide the target cueing so that the bomber would not need to disclose its position by activating its radar. The missiles themselves could potentially be based on existing airframes, such as those of the Patriot MIM-104 or SM-2ER RIM-67 (which would have significantly longer ranges when air launched instead of surface launched), perhaps coupled with a small second stage for the terminal engagement.[18]

---

missiles that China is acquiring will force U.S. carrier-based aircraft, which, in any case, will be available in fewer numbers than land-based fighters, to operate from similar distances. See Cliff, Burles, et al., 2007, pp. 71–76, 89–93.

[18] "Boeing Integrated Defense Systems: Boeing (Rockwell) B-1B Lancer," *Jane's Aircraft Upgrades*, February 1, 2008; "MIM-104 Patriot," *Jane's Strategic Weapon Systems*, March 31, 2008; "RIM-66/-67/-156 Standard SM-1/-2, RIM-161 Standard SM-3, and SM-6," *Jane's*

In addition to improving its capabilities to defend Taiwan's airspace, the USAF should also examine ways to improve its capabilities to conduct offensive operations against China. Although China's air defenses are formidable and growing ever more so, the difficulties associated with defending Taiwan's airspace may be such that the most effective way to defeat China's air force in a conflict over Taiwan would be to attack China's aircraft while they were on the ground. If Chinese attacks on U.S. air bases on Okinawa force U.S. land-based aircraft to operate from more-distant bases, therefore, the USAF should consider launching attacks on China's air bases as well.[19]

As noted in Chapter Ten, China has devoted significant effort to sheltering its fighter aircraft, but it is nonetheless possible to destroy sheltered aircraft with PGMs. Moreover, as is the case with the United States, China does not have shelters for its large, high-value aircraft, such as bombers, airborne warning and control, and EW aircraft. The challenge would be to deliver munitions against these targets in the face of China's highly capable, long-range SAMs and other air defenses. The USAF's stealthy B-2 bomber can potentially penetrate those defenses, but it is an extremely valuable aircraft that U.S. commanders might not be willing to risk on that mission. And even if they were, the number of B-2s is small (20), so that, although each B-2 can carry 20 GBU-31 2,000-lb Joint Direct Attack Munitions (JDAMs), it could take them a relatively long time to destroy a significant number of China's aircraft shelters.[20]

If a new-generation bomber becomes available, it will be able to augment the capability currently provided by the B-2. An alternative to bombers penetrating into China's territory, however, would be a long-range, stealthy cruise missile that could be launched at standoff

---

Strategic Weapon Systems, March 31, 2008. We are grateful to colleague Eric Gons for these suggestions.

[19] Attacks on targets other than airfields could be regarded as escalatory by Beijing and, given that China is a nuclear power, should probably be avoided or conducted only after careful consideration of the potential conflict escalation that could result.

[20] IISS, 2008, p. 37; "Northrop Grumman (Northrop) B-2A Spirit," *Jane's Aircraft Upgrades*, April 28, 2008.

ranges from bombers that the USAF possesses in larger numbers than the B-2. The stealthy Joint Air-to-Surface Standoff Missile–Extended Range (JASSM-ER) launched from B-1s might be able to play this role. The JASSM-ER will have a range of more than 500 nm, and each B-1 can carry 24 JASSMs. (The USAF has a total of approximately 95 B-1s, although only about 65 are currently maintained at combat readiness.) In the future, China may acquire from Russia the S-400 SAM system, which has an engagement range of about 200 nm, so the B-1 might need to launch its weapons at least 200 nm from China's land borders. With JASSM-ER, however, it would still be able to reach Chinese air-fields 300 nm or more inland. This would force China's short-range fighters, such as the J-10, J-8, and J-7, to risk being destroyed on the ground or to operate from bases deeper inland, where their combat effectiveness would be significantly reduced. In other words, China could be presented with a geographic challenge similar to the one that the USAF would face, partially leveling the playing field.[21]

To reach targets further inland, a missile like the Advanced Cruise Missile (ACM) could potentially be used. The ACM is stealthy and has a range of 1,865 miles, so it could it reach all airfields of interest in China even when launched from a safe standoff range by the B-52, which is its current launch platform. Each of the USAF's 85 current combat-ready B-52s can carry up to 12 ACMs at a time, although the total inventory of ACMs, which are no longer in production, is only 450. In 2007, the USAF announced that it was going to withdraw all ACMs from service by 2012. If it were possible to instead refurbish and convert these weapons to carry conventional warheads (under an earlier planned service-life extension program, they were to have remained in service until 2030, and, under the original acquisition program, some were planned to be conventionally armed), they would provide the USAF with a conventional standoff strike weapon capable of attack-

---

[21] "AGM-158A JASSM (Joint Air-to-Surface Standoff Missile), AGM-158B JASSM-ER," *Jane's Air-Launched Weapons*, January 23, 2008; IISS, 2008, p. 37; "S-400 Triumf (SA-21 'Growler')," *Jane's Strategic Weapon Systems*, February 21, 2008.

ing targets deep in China's interior.[22] Alternatively, a new penetrating missile with a range significantly greater than that of the JASSM-ER could be developed.

In addition to direct physical attacks on China, the United States should also explore other means of degrading China's military operations. These could include computer network attack, EW, and other types of information operations.[23]

Implications for Taiwan

In a conflict over Taiwan, the capabilities of Taiwan's armed forces would be critical to the outcome, even if the United States intervened on a large scale. It is important, therefore, to assess the implications of China's air force employment concepts and capabilities not just for the United States but also for Taiwan.

First, as with U.S. forces in the western Pacific, since a PRC use of force against Taiwan could develop with little warning, even in the absence of an obvious crisis with the PRC, Taiwan should take steps to prevent China from collecting information on military and sensitive civilian information systems or on Taiwan's early warning, command-and-control, SAM, and other sensors and communication systems. Moreover, in the event that an attack was planned, Taiwan's intelligence collectors should expect extensive efforts to deceive them about the locations and posture of Chinese forces. Taiwan's military should also ensure, to the maximum extent practical, that its information systems are protected from network intrusions, some of which may be going on today, or denial-of-service attacks. It should also plan and train for the possibility that some of these systems would fail or be compromised in a conflict with the PRC. And, once a Chinese offensive air campaign is under way, Taiwan should be prepared to deal with massive electronic jamming.

---

[22] "AGM-129 Advanced Cruise Missile (ACM)," *Jane's Air-Launched Weapons*, March 28, 2007; IISS, 2008, p. 37.

[23] We are grateful to Michael Chase of the U.S. Naval War College for this observation.

It is clearly not feasible for Taiwan to acquire enough missile defense systems to protect it against the simultaneous arrival of 100 or more ballistic missile warheads; therefore, it should accept the likelihood that the runways of most of its military airfields would be rendered at least temporarily unusable and that many or most unsheltered aircraft would be damaged or destroyed in the initial salvo of ballistic missiles. Nonetheless, additional missile defenses, such as the six PAC-3 batteries that Taiwan plans to acquire, will have some utility by increasing the number of ballistic missiles that China would have to launch to be certain of putting out of action the runways at all of Taiwan's military airfields. If they were concentrated near one or two randomly chosen air bases, moreover, Taiwan's PAC-3 and PAC-2 systems might be able to keep at least some runways usable. For them to be effective, however, they must be relocated on a regular basis to unpredictable locations. Otherwise, they are unlikely to survive China's initial ballistic missile salvo.

At least as important as, and possibly more cost-effective than, active missile defenses would be passive defense measures, such as building shelters to protect Taiwan's combat aircraft from ballistic missile attack; hardening runways and fuel and repair facilities; and increasing rapid runway repair capabilities at Taiwan's air bases. Aircraft shelters being far less expensive than aircraft, the number of shelters would ideally be several times the number of Taiwan's combat aircraft, with each aircraft randomly assigned to one of several different shelters every time it returned to base. This would significantly increase the number of aircraft that would survive China's cruise missile and aircraft attacks, particularly after a ballistic missile salvo, when runways would be unusable until repaired and the aircraft would be unable to get aloft. Mobile point-defense systems, such as SLAMRAAM, could help defend Taiwan's air bases and other key targets against attacks by aircraft, cruise missiles, and PGMs, further increasing the number that would survive these attacks. Finally, even if hostilities have not actually occurred, if there are indications that China might use force against Taiwan, as many aircraft as possible should be maintained aloft, to ensure that at least some would be available after an initial ballistic missile attack to engage the Chinese cruise missiles and aircraft that would follow.

Taiwan's defenders should be prepared for the PRC's cruise missiles and aircraft to approach Taiwan not on a direct line from their launch points but from all directions. The attacking aircraft and missiles should be expected to focus their attacks first on Taiwan's own air and missile capabilities. An airborne landing, if attempted, would most likely occur in a lightly defended location in an area where the PRC could ensure continuous air superiority between the point of embarkation and the landing zone.

Taiwan should also expect attacks on government, water, and electric installations and, if a prolonged campaign is expected, on key economic targets. Although it is not possible to defend all such targets, mitigating actions can be taken, such as ensuring that backup installations exist and evacuating government facilities if there are indications that China might use force against Taiwan.

Maintaining viable combat capabilities in the face of PRC air and missile attacks will be increasingly challenging for Taiwan as the PLA's capabilities improve but is nonetheless feasible if systematic, sustained, and carefully chosen investments are made. The longer Taiwan is able to deny the PRC air superiority over Taiwan, the more combat power the United States will be able to bring to the defense of Taiwan and the better the chances of a successful defense of the island.

Bibliography

"AGM-129 Advanced Cruise Missile (ACM)," *Jane's Air-Launched Weapons*, March 28, 2007.

"AGM-158A JASSM (Joint Air-to-Surface Standoff Missile), AGM-158B JASSM-ER," *Jane's Air-Launched Weapons*, January 23, 2008.

Air Force Dictionary 《空军大辞典》, Shanghai: 上海辞书出版社 [Shanghai Dictionary Press], 1996.

"Air Power Australia," as of December 19, 2009, update. As of December 29, 2009:
http://www.ausairpower.net/

Allen, Kenneth W., *People's Republic of China, People's Liberation Army Air Force*, Washington, D.C.: Defense Intelligence Agency, DIC-1300-445-91, April 15, 1991.

———, "PLAAF Modernization: An Assessment," in James R. Lilley and Chuck Downs, eds., *Crisis in the Taiwan Strait*, Washington, D.C.: National Defense University Press, September 1997, pp. 217–248. As of December 29, 2009:
http://purl.access.gpo.gov/GPO/LPS51431

———, "PLA Air Force Organization," in James C. Mulvenon and Andrew N. D. Yang, eds., *The People's Liberation Army as Organization: Reference Volume v1.0*, Santa Monica, Calif.: RAND Corporation, CF-182-NSRD, 2002, pp. 346–457. As of December 29, 2009:
http://www.rand.org/pubs/conf_proceedings/CF182/

———, "PLA Air Force, 1949–2002: Overview and Lessons Learned," in Laurie Burkitt, Andrew Scobell, and Larry M. Wortzel, eds., *The Lessons of History: The Chinese People's Liberation Army at 75*, Carlisle, Pa.: Strategic Studies Institute, July 2003, pp. 89–156. As of December 29, 2009:
http://www.carlisle.army.mil/ssi/pubs/2003/pla75/pla75.pdf

———, "Reforms in the PLA Air Force," *China Brief*, Vol. 5, No. 15, July 5, 2005a. As of December 29, 2009:
http://www.jamestown.org/single/?no_cache=1&tx_ttnews[tt_news]=3875

————, "The PLA Air Force: 2006–2010," paper presented at the CAPS-RAND-CEIP International Conference on PLA Affairs, Taipei, November 10–12, 2005b.

Allen, Kenneth W., and Maryanne Kivlehan-Wise, "Implementing PLA Second Artillery Doctrinal Reforms," in James C. Mulvenon and David Michael Finkelstein, eds., *China's Revolution in Doctrinal Affairs: Emerging Trends in the Operational Art of the Chinese People's Liberation Army*, Alexandria, Va.: CNA Corporation, December 2005, pp. 159–200.

Allen, Kenneth W., Glenn Krumel, and Jonathan D. Pollack, *China's Air Force Enters the 21st Century*, Santa Monica, Calif.: RAND Corporation, MR-580-AF, 1995. As of December 29, 2009:
http://www.rand.org/pubs/monograph_reports/MR580/

"Almaz/Antei Concern of Air Defence S-75 Family of (SA-2 'Guideline') Low- to High-Altitude Surface-to-Air Missile Systems," *Jane's Land-Based Air Defence*, September 16, 2004.

Bi Xinglin [薛兴林], ed.,《战役理论学习指南》[*Campaign Theory Study Guide*], Beijing: 国防大学出版社 [National Defense University Press], 2002.

Blasko, Dennis J., "PLA Ground Forces: Moving Toward a Smaller, More Rapidly Deployable, Modern Combined Arms Force," in James C. Mulvenon and Andrew N. D. Yang, eds., *The People's Liberation Army as Organization*, Santa Monica, Calif.: RAND Corporation, CF-182-NSRD, 2002, pp. 309–345. As of December 29, 2009:
http://www.rand.org/pubs/conf_proceedings/CF182/

————, *The Chinese Army Today: Tradition and Transformation for the 21st Century*, London: Routledge, 2006.

"Boeing Integrated Defense Systems: Boeing (Rockwell) B-1B Lancer," *Jane's Aircraft Upgrades*, February 1, 2008.

Burles, Mark, *Chinese Policy Toward Russia and the Central Asian Republics*, Santa Monica, Calif.: RAND Corporation, MR-1045-AF, 1999. As of December 29, 2009:
http://www.rand.org/pubs/monograph_reports/MR1045/

Burles, Mark, and Abram N. Shulsky, *Patterns in China's Use of Force: Evidence from History and Doctrinal Writings*, Santa Monica, Calif.: RAND Corporation, MR-1160-AF, 2000. As of December 29, 2009:
http://www.rand.org/pubs/monograph_reports/MR1160/

Byman, Daniel, and Roger Cliff, *China's Arms Sales: Motivations and Implications*, Santa Monica, Calif.: RAND Corporation, MR-1119-AF, 1999. As of December 29, 2009:
http://www.rand.org/pubs/monograph_reports/MR1119/

"CAC J-10," *Jane's All the World's Aircraft*, April 14, 2010.

Cai Fengzhen [蔡凤震], and Tian Anping [田安平], eds., 《空天战场与中国空军》 [*Air and Space Battlefield and China's Air Force*], Beijing: 解放军出版社 [Liberation Army Press], 2004.

Cai Fengzhen [蔡凤震], Tian Anping [田安平], Chen Jiesheng [陈杰生], Cheng Jian [程建], Zheng Dongliang [郑东良], Liang Xiaoan [梁小安], Deng Pan [邓攀], and Guan Hua [管桦], eds., 《空天一体作战学》 [*The Study of Integrated Air and Space Operations*], Beijing: 解放军出版社 [Liberation Army Press], 2006.

Central People's Government of the People's Republic of China, 〈中华人民共和国中央军事委员会〉 ["The Central Military Commission of the People's Republic of China"], web page, March 15, 2008. As of July 13, 2009, in Chinese: http://www.gov.cn/test/2008-03/15/content_921057.htm

Cheng Li, "China's Midterm Jockeying: Gearing Up for 2012 (Part 3: Military Leaders)," *China Leadership Monitor*, No. 33, June 28, 2010, p. 2.

"China: Air Force," *Jane's World Air Forces*, February 23, 2007.

"China: Air Force," *Jane's Sentinel Security Assessment: China and Northeast Asia*, February 4, 2008.

China Naval Encyclopedia 《中国海军百科全书》, Beijing: 海潮出版社 [Haichao Press], 1999.

"China: PLA Publishes New List of Senior Military Leaders—Report from the Beijing News Center by Yang Fan: 'PLA Publishes List of Senior Officers of the Four General Departments,'" *Wen Wei Po*, January 4, 2006.

"Chinese Electronic Warfare (EW) Aircraft," *Jane's Electronic Mission Aircraft*, August 30, 2008.

"Chinese Laser-Guided Bombs (LGBs)," *Jane's Air-Launched Weapons*, January 17, 2008.

"Chinese Signals Intelligence (SIGINT) Air Vehicles," *Jane's Electronic Mission Aircraft*, January 8, 2007.

Cliff, Roger, *The Military Potential of China's Commercial Technology*, Santa Monica, Calif.: RAND Corporation, MR-1292-AF, 2001. As of December 29, 2009: http://www.rand.org/pubs/monograph_reports/MR1292/

Cliff, Roger, Mark Burles, Michael S. Chase, Derek Eaton, and Kevin L. Pollpeter, *Entering the Dragon's Lair: Chinese Antiaccess Strategies and Their Implications for the United States*, Santa Monica, Calif.: RAND Corporation, MG-524-AF, 2007. As of December 29, 2009: http://www.rand.org/pubs/monographs/MG524/

Cliff, Roger, and David A. Shlapak, *U.S.-China Relations After Resolution of Taiwan's Status*, Santa Monica, Calif.: RAND Corporation, MG-567-AF, 2007. As of December 29, 2009:
http://www.rand.org/pubs/monographs/MG567/

Cole, Bernard D., *The Great Wall at Sea: China's Navy Enters the Twenty-First Century*, Annapolis, Md.: Naval Institute Press, 2001.

Crane, Keith, Roger Cliff, Evan S. Medeiros, James C. Mulvenon, and William H. Overholt, *Modernizing China's Military: Opportunities and Constraints*, Santa Monica, Calif.: RAND Corporation, MG-260-1-AF, 2005. As of December 29, 2009:
http://www.rand.org/pubs/monographs/MG260-1/

Cui Changqi [崔长崎], Ji Rongren [纪荣仁], Min Zengfu [闵增富], Yuan Jingwei [袁静伟], Hu Siyuan [胡思远], Tian Tongshun [田同顺], Ruan Guangfeng [阮光峰], Hong Baocai [洪宝才], Meng Qingquan [孟庆全], Cao Xiumin [曹秀敏], Dai Jianjun [戴建军], Han Jibing [韩继兵], Wang Jicheng [王冀城], and Wang Xuejin [王学进], 《21世纪初空袭与反空袭》 [*Air Raids and Counter–Air Raids in the Early 21st Century*], Beijing: 解放军出版社 [Liberation Army Press], 2002.

Declaration Concerning the Laws of Naval War, 208 Consol. T.S. 338, 1909.

"DF-11 (CSS-7/M-11)," *Jane's Strategic Weapon Systems*, June 18, 2007.

"DF-15 (CSS-6/M-9)," *Jane's Strategic Weapon Systems*, July 6, 2007.

"DF-21 (CSS-5)," *Jane's Strategic Weapon Systems*, June 18, 2007.

"DF-25," *Jane's Strategic Weapon Systems*, June 29, 2007.

Downs, Erica Strecker, *China's Quest for Energy Security*, Santa Monica, Calif.: RAND Corporation, MR-1244-AF, 2000. As of December 29, 2009:
http://www.rand.org/pubs/monograph_reports/MR1244/

"Fei Teng Guided Bombs (FT-1, FT-2, FT-3, FT-5)," *Jane's Air-Launched Weapons*, January 13, 2010.

Fisher, Richard D., Jr., "PLA Air Force Equipment Trends," in Stephen J. Flanagan and Michael E. Marti, eds., *The People's Liberation Army and China in Transition*, Washington, D.C.: National Defense University Press, August 2003, pp. 139–176. As of December 29, 2009:
http://purl.access.gpo.gov/GPO/LPS36357

Fulghum, David A., and Douglas Barrie, "Non-War: China Accelerates Focus on Disruption, Asymmetric Tactics," *Aviation Week and Space Technology*, March 10, 2008.

Ge Xinqing [葛信卿], 《导弹作战研究》 [*Research on Missile Operations*], Beijing: 解放军出版社 [Liberation Army Press], 2005.

Gill, Bates, James Mulvenon, and Mark A. Stokes, "The Chinese Second Artillery Corps: Transition to Credible Deterrence," in James C. Mulvenon and Andrew N. D. Yang, eds., *The People's Liberation Army as Organization: Reference Volume v.1.0*, Santa Monica, Calif.: RAND Corporation, CF-182-NSRD, 2002, pp. 510–586. As of December 29, 2009:
http://www.rand.org/pubs/conf_proceedings/CF182/

"HQ-9/-15, and RF-9 (HHQ-9 and S-300)," *Jane's Strategic Weapon Systems*, January 28, 2008.

"HUMRAAM/SLAMRAAM," *Jane's Land-Based Air Defence*, October 22, 2008.

"IAI Harpy and Cutlass," *Jane's Unmanned Aerial Vehicles and Targets*, March 25, 2008.

IISS—*See* International Institute for Strategic Studies.

Information Office of the State Council of the People's Republic of China, *China's National Defense in 2002*, Beijing: New Star Publishers, 2002.

———, *China's National Defense in 2004*, Beijing, December 27, 2004. As of December 29, 2009:
http://english.people.com.cn/whitepaper/defense2004/defense2004.html

———, *China's National Defense in 2006*, Beijing, December 2006. As of July 13, 2007:
http://english.people.com.cn/whitepaper/defense2006/defense2006.html

———, *China's National Defense in 2008*, Beijing, January 21, 2009. As of February 10, 2009:
http://www.china.org.cn/government/whitepaper/node_7060059.htm

International Institute for Strategic Studies, *The Military Balance 1999/2000*, London: Oxford University Press, 1999.

———, *The Military Balance 2007*, London: Routledge, 2007.

———, *The Military Balance 2008*, London: Routledge, 2008.

———, *The Military Balance 2010*, London: Routledge, 2010.

Joint Staff, "Joint Electronic Library," as of December 10, 2009, update. As of December 29, 2009:
http://www.dtic.mil/doctrine/

"KD-63 (Kong Di-63)," *Jane's Air-Launched Weapons*, January 25, 2008.

Khalilzad, Zalmay, David T. Orletsky, Jonathan D. Pollack, Kevin L. Pollpeter, Angel Rabasa, David A. Shlapak, Abram N. Shulsky, and Ashley J. Tellis, *The United States and Asia: Toward a New U.S. Strategy and Force Posture*, Santa Monica, Calif.: RAND Corporation, MR-1315-AF, 2001. As of December 29, 2009:
http://www.rand.org/pubs/monograph_reports/MR1315/

Khalilzad, Zalmay, Abram N. Shulsky, Daniel Byman, Roger Cliff, David T. Orletsky, David A. Shlapak, and Ashley J. Tellis, *The United States and a Rising China: Strategic and Military Implications*, Santa Monica, Calif.: RAND Corporation, MR-1082-AF, 1999. As of December 29, 2009: http://www.rand.org/pubs/monograph_reports/MR1082/

Kopp, Carlo, *Sukhoi Flankers: The Shifting Balance of Regional Air Power*, Air Power Australia technical report APA-TR-2007-0101, January 2007, updated September 2009. As of April 18, 2008: http://www.ausairpower.net/APA-Flanker.html

———, *XAC (Xian) H-6 Badger*, Air Power Australia technical report APA-TR-2007-0705, July 2007. As of January 11, 2008: http://www.ausairpower.net/APA-Badger.html

Lanzit, Kevin, and Kenneth W. Allen, "Right-Sizing the PLA Air Force: New Operational Concepts Define a Smaller, More Capable Force," in Roy Kamphausen and Andrew Scobell, eds., *Right-Sizing the People's Liberation Army: Exploring the Contours of China's Military*, Carlisle, Pa.: Strategic Studies Institute, U.S. Army War College, September 2007, pp. 437–478. As of December 29, 2009: http://purl.access.gpo.gov/GPO/LPS85666

Lewis, John W., and Xue Litai, "China's Search for a Modern Air Force," *International Security*, Vol. 24, No. 1, Summer 1999, pp. 64–94.

Li Daguang [李大光], 《太空战》 [*Space War*], Beijing: 军事科学出版社 [Military Science Press], 2001.

Li Rongchang [李荣常] and Cheng Jian [程建], 《空天一体信息作战》 [*Integrated Air and Space Information Warfare*], Beijing: 军事科学出版社 [Military Science Press], 2003.

Liu Yazhou [刘亚洲], Qiao Liang [乔良], and Wang Xiangsui [王湘穗], 〈战争空中化与中国空军〉 ["Combat in the Air and China's Air Force"], in Shen Weiguang [沈伟光], ed., Xie Xizhang [解玺璋] and Ma Yaxi [马亚西], assoc. eds., 《中国军事变革》 [*China's Military Transformation*], 新华出版社 [Xinhua Press], 2003, pp. 79–104.

"LS-6 Glide Bomb," *Jane's Air-Launched Weapons*, January 17, 2008.

"LT-2 Laser Guided Bomb," *Jane's Air-Launched Weapons*, July 27, 2007.

Lu Lihua [芦利华], 《军队指挥理论学习指南》 [*Military Command Theory Study Guide*], Beijing: 国防大学出版社 [National Defense University Press], 2004.

Medeiros, Evan S., *China's International Behavior: Activism, Opportunism, and Diversification*, Santa Monica, Calif.: RAND Corporation, MG-850-AF, 2009. As of September 24, 2010: http://www.rand.org/pubs/monographs/MG850/

Medeiros, Evan S., Roger Cliff, Keith Crane, and James C. Mulvenon, *A New Direction for China's Defense Industry*, Santa Monica, Calif.: RAND Corporation, MG-334-AF, 2005. As of December 29, 2009:
http://www.rand.org/pubs/monographs/MG334/

Medeiros, Evan S., Keith Crane, Eric Heginbotham, Norman D. Levin, Julia F. Lowell, Angel Rabasa, and Somi Seong, *Pacific Currents: The Responses of U.S. Allies and Security Partners in East Asia to China's Rise*, Santa Monica, Calif.: RAND Corporation, MG-736-AF, 2008. As of December 29, 2009:
http://www.rand.org/pubs/monographs/MG736/

"MIM-104 Patriot," *Jane's Strategic Weapon Systems*, March 31, 2008.

Mulvenon, James C., and David Michael Finkelstein, eds., *China's Revolution in Doctrinal Affairs: Emerging Trends in the Operational Art of the Chinese People's Liberation Army*, Alexandria, Va.: CNA Corporation, December 2005.

"National Air Intelligence Radar Network Realizes Complete Early Warning Coverage" 〈我国空情雷达网实现全域预警覆盖〉,《航空知识》[*Aerospace Knowledge*], No. 12, December 2007, p. 8.

"Northrop Grumman (Northrop) B-2A Spirit," *Jane's Aircraft Upgrades*, April 28, 2008.

Office of Naval Intelligence, *China's Navy 2007*, Washington, D.C.: Department of the Navy, 2007.

Office of the Secretary of Defense, *Annual Report to Congress: Military Power of the People's Republic of China 2006*, Washington D.C.: U.S. Department of Defense, 2006. As of December 29, 2009:
http://handle.dtic.mil/100.2/ADA449718

———, *Annual Report to Congress: Military Power of the People's Republic of China 2007*, Washington, D.C.: U.S. Department of Defense, 2007. As of September 27, 2010:
http://www.defense.gov/pubs/pdfs/070523-China-Military-Power-final.pdf

———, *Annual Report to Congress: Military Power of the People's Republic of China 2008*, Washington D.C.: U.S. Department of Defense, 2008. As of December 29, 2009:
http://handle.dtic.mil/100.2/ADA477533

———, *Annual Report to Congress: Military Power of the People's Republic of China 2009*, Washington D.C.: U.S. Department of Defense, 2009. As of July 13, 2009:
http://www.defenselink.mil/pubs/pdfs/China_Military_Power_Report_2009.pdf

———, *Annual Report to Congress: Military and Security Developments Involving the People's Republic of China 2010*, Washington, D.C.: U.S. Department of Defense, 2010. As of September 7, 2010:
http://www.defense.gov/pubs/pdfs/2010_CMPR_Final.pdf

Peng Xiwen [彭希文] and Bi Xinglin [薛兴林], 《空袭与反空袭怎样打》 [*How Air Attack and Air Defense Are Fought*], 中国青年出版社 [Chinese Youth Press], 2001.

People's Liberation Army Air Force [中国人民解放军空军], 《空军战术学》 [*Study of Air Force Tactics*], Beijing: 解放军出版社 [Liberation Army Press], 1994.

———, 《中国空军百科全书》 [*China Air Force Encyclopedia*], Beijing: 航空工业出版社 [Aviation Industry Press], 2005.

"People's Liberation Army Air Force," GlobalSecurity.org, as of April 27, 2005, modification. As of December 27, 2007: http://www.globalsecurity.org/military/world/china/plaaf-intro.htm

PLAAF—*See* People's Liberation Army Air Force.

Pollpeter, Kevin L., *U.S.-China Security Management: Assessing the Military-to-Military Relationship*, Santa Monica, Calif.: RAND Corporation, MG-143-AF, 2004. As of December 29, 2009: http://www.rand.org/pubs/monographs/MG143/

"PRC General Xu Qiliang's Promotion to Spur Combined Operations of Armed Forces," *The Standard*, July 14, 2004.

"RIM-66/-67/-156 Standard SM-1/-2, RIM-161 Standard SM-3, and SM-6," *Jane's Strategic Weapon Systems*, March 31, 2008.

"S-400 Triumf (SA-21 'Growler')," *Jane's Strategic Weapon Systems*, February 21, 2008.

"SA-10/20 'Grumble' (S-300, S-300 PMU, Buk/Favorit/5V55/48N6," *Jane's Strategic Weapon Systems*, December 29, 2006.

"SAC J-8," *Jane's All the World's Aircraft*, March 9, 2009.

"SAC (Sukhoi Su-27) J-11B," *Jane's All the World's Aircraft*, November 12, 2009.

"SAC Y-8/Y-9 (Special Mission Versions)," *Jane's All the World's Aircraft*, November 12, 2009.

"SD-10 (PL-12)," *Jane's Air-Launched Weapons*, January 22, 2008.

Shulsky, Abram N., *Deterrence Theory and Chinese Behavior*, Santa Monica, Calif.: RAND Corporation, MR-1161-AF, 2000. As of December 29, 2009: http://www.rand.org/pubs/monograph_reports/MR1161/

"SinoDefence.com," undated home page. As of December 29, 2009: http://www.sinodefence.com/

Sokolsky, Richard, Angel Rabasa, and C. Richard Neu, *The Role of Southeast Asia in U.S. Strategy Toward China*, Santa Monica, Calif.: RAND Corporation, MR-1170-AF, 2001. As of December 29, 2009: http://www.rand.org/pubs/monograph_reports/MR1170/

Stillion, John, and David T. Orletsky, *Airbase Vulnerability to Conventional Cruise-Missile and Ballistic-Missile Attacks: Technology, Scenarios, and U.S. Air Force Responses*, Santa Monica, Calif.: RAND Corporation, MR-1028-AF, 1999. As of December 29, 2009:
http://www.rand.org/pubs/monograph_reports/MR1028/

Stokes, Mark A., "The Chinese Joint Aerospace Campaign: Strategy, Doctrine, and Force Modernization," in James C. Mulvenon and David Michael Finkelstein, eds., *China's Revolution in Doctrinal Affairs: Emerging Trends in the Operational Art of the Chinese People's Liberation Army*, Alexandria, Va.: CNA Corporation, 2005, pp. 221–306.

"Sukhoi Su-27," *Jane's All the World's Aircraft*, February 14, 2008.

"Sukhoi Su-30 (Su-27PU)," *Jane's All the World's Aircraft*, February 14, 2008.

Swaine, Michael D., and Ashley J. Tellis, *Interpreting China's Grand Strategy: Past, Present, and Future*, Santa Monica, Calif.: RAND Corporation, MR-1121-AF, 2000. As of December 29, 2009:
http://www.rand.org/pubs/monograph_reports/MR1121/

"Tien Kung 1/2/3 (Sky Bow)," *Jane's Strategic Weapon Systems*, February 22, 2008.

"TKP: Ma Xiaotian to Become President of PLA National Defense University—Report by Wu Yue: 'Ma Xiaotian Promoted to Office of President of Defense University,'" *Ta Kung Pao*, August 18, 2006.

"Tor," *Jane's Land-Based Air Defence*, April 13, 2010.

U.S. Air Force, *Air Force Basic Doctrine*, Washington, D.C., Air Force Doctrine Document 1, November 17, 2003. As of December 29, 2009:
http://purl.access.gpo.gov/GPO/LPS88904

Wang Fengshan [王凤山], Yang Jianjun [杨建军], and Chen Jiesheng [陈杰生], eds.,《信息时代的国家防空》[*National Air Defense in the Information Age*], Beijing: 航空工业出版社 [Aviation Industry Press], 2004.

Wang Houqing [王厚卿] and Zhang Xingye [张兴业], eds.,《战役学》[*Study of Campaigns*], Beijing: 国防大学出版社 [National Defense University Press], 2000.

Wuxi air defense website, undated. As of September 24, 2010:
http://www.xchmf.gov.cn/xpe/portal/2aa77a1c-114a-1000-8439-d31db1c79082?
objectId=oid:95bbf14-1142-1000-8873-e25d6dbdf21b&uuid=
95bbf14-1142-1000-8874-e25d6dbdf21b

"Xian Aircraft Company: XAC JH-7," *Jane's All the World's Aircraft*, August 2, 2007.

"Xian Aircraft Industries Group: XAC H-6," *Jane's All the World's Aircraft*, August 2, 2007.

Yang Xuejun [杨学军] and Zhang Wangxin [张望新], eds.,《优势来自空间: 论空间战场与空间作战》[*Advantage Comes from Space: On the Space Battlefield and Space Operations*], Beijing: 国防工业出版社 [National Defense Industry Press], 2006.

"YJ-91, KR-1 (Kh-31P)," *Jane's Air-Launched Weapons*, October 15, 2007.

You Ji, "Sorting Out the Myths About Political Commissars," in Nan Li, ed., *Chinese Civil-Military Relations: The Transformation of the People's Liberation Army*, New York: Routledge, 2006, pp. 89–116.

Zhan Xuexi [展学习], ed.,《战役学研究》[*Campaign Studies Research*], Beijing: 国防大学出版社 [National Defense University Press], 1997.

Zhang Xiaoming, "Air Combat for the People's Republic: The People's Liberation Army Air Force in Action, 1949–1969," in Mark A. Ryan, David Michael Finkelstein, and Michael McDevitt, eds., *Chinese Warfighting: The PLA Experience Since 1949*, Armonk, N.Y.: M. E. Sharpe, 2003, pp. 270–300.

Zhang Yuliang [张玉良], ed., 《战役学》 [*Study of Campaigns*], Beijing: 国防大学出版社 [National Defense University Press], 2006.

Zhang Zhiwei [张志伟] and Feng Chuanjiang [冯传奖], 〈试析未来空天一体作战〉["Thoughts on Future Integrated Air-Space Operations"],《军事科学》[*Military Science*], Vol. 2, 2006, pp. 52–59.

Index